The Alaska Geographic Society

The Alaska Geographic Society is a nonprofit organization — exploring new frontiers of knowledge across the lands of the polar rim, learning how other men and other countries live in their "Norths," putting the geography book back in the classroom, exploring new methods of teaching and learning — sharing in the excitement of discovery in man's wonderful new world North of 51°16′.

Members of **The Alaska Geographic Society** receive **ALASKA GEOGRAPHIC**, a quality magazine in color which devotes each quarterly issue to monographic in-depth coverage of a northern geographic region or resource-oriented subject.

The Alaska Geographic Society membership dues are $39 per year; $43 to non-U.S. addresses (80 percent of this amount is for a 1-year subscription to **ALASKA GEOGRAPHIC**).

Box 93370, Anchorage, Alaska 99509

Additional copies of this issue and back issues of ALASKA GEOGRAPHIC are available to nonmembers as well as members. Write to The Alaska Geographic Society, Box 93370, Anchorage, Alaska 99509.

Cover: *Fishermen lifting an Italian floating trap off live bit in Chignik Bay, Alaska, about 1911-12. There were 42 Italians employed on 1 trap, working in 2 shifts. Lead of trap was 3 miles long. (San Francisco Maritime Museum, Alaska Packers' Association Collection)*

To teach many more to better know and use our natural resources

ALASKA GEOGRAPHIC

Published quarterly by The Alaska Geographic Society, Box 93370, Anchorage, Alaska 99509

VOL. 3, NO. 4,

IN THIS ISSUE:

An album of historical photos–

THE SILVER YEARS

of the Alaska Canned Salmon Industry

COLLECTED AND EDITED
BY LAURENCE FREEBURN

Library of Congress Catalog Card Number: 72-92087
ISSN 0361-1353 Key title: Alaska Geographic
International Standard Book Number: 0-88240 082-7

Editors: Robert A. Henning, C. H. Rosenthal, Barbara Olds, Ed Reading
Designed by Roselyn Pape

Second-class postage paid at Anchorage, AK 99509.

MATERIAL SOUGHT—The editors seek a wide variety of informative material on the lands north of 51°16′ on geographic subjects—anything to do with resources and their uses, with heavy emphasis on quality color photography, from Alaska, Northern Canada, Siberia, Japan—all geographic areas that have a relationship to Alaska in a physical or economic sense. We do not want material done in excessive scientific terminology. A query to the editors is suggested. Modest payments for all material upon publication.

Printed in Thailand.

Postmaster: Send form 3579 to Alaska Geographic, Box 93370, Anchorage, AK 99509.

An album of historical photos—

THE SILVER YEARS of the Alaska Canned Salmon Industry

COLLECTED AND EDITED
BY LAURENCE FREEBURN

Right—*Sunday, no fishing, clean laundry on a forestay, nets to be mended and time for a post-mortem on last week's fishing over a mug of coffee.*

JOHN N. COBB COLLECTION, UNIVERSITY OF WASHINGTON LIBRARY

CONTENTS

Hand-by-hand loading of wooden cases aboard the Star of Alaska *in the Bering Sea earns these fishermen run money before they return home.*

FRED HEICK COLLECTION, SAN FRANCISCO MARITIME MUSEUM

FOREWORD

The business of putting up Alaskan salmon probably should be traced back to the early migrants through the land from Asia across the Bering land bridge (bridges?), slowly moving bands resting here and there as they marched or paddled from one place to another gradually passing to lands south or settling in at or near the mouths of countless streams giving into the Pacific Ocean, streams frequently crowded with seemingly endless processions of salmon going to their spawning grounds. It was simple to pile stones across the shallow waters of a stream pouring across the tide flats at low water. At high water the salmon came in from the sea and when the tide turned and dropped, they were left corralled and flopping on shallow ground above the rock barriers. Thus may have been born Alaska's first salmon trap.

And the people of those times, catching more than they could use at the moment, prepared the excess salmon for later eating, by drying, by smoking, by burying in pits. These, then, were the first salmon packers—and in times of plenty, they had salmon left over for trade with other Natives less fortunate—and thus was born the industry, the catching and processing of salmon for profit.

In the mid-19th century (in Alaska an indeterminate time before the purchase from Russia in 1867) the business of salting salmon at various locations became a standard practice. Materials for barrels, tools for coopering, and salt were transported to fishing locations near the teeming streams. Soon, the salting became canning, with tin plate, solder, salt and cases shipped to the old saltery locations. Then came the Chinese crews, the Filipino crews, the stake traps in the salmon streams themselves, and fish wheels, and beach seines drawn in by great steam donkeys, canneries, shore stations, floaters, ships and men and cans and competition everywhere—the financial centers of San Francisco and Seattle and other Northwest ports buzzing with the fever of the taking and the profiting—then the great pile traps, the floating traps, the "Iron Chink," the high-speed line—and the declining years of fewer salmon, higher prices, no Chinese crews, shorter seasons, the bigger and bigger seine boats with faster and faster catching gear, the end of sail in the picturesque Bristol Bay gill-netters' fleets, the advent of the freezer ship and far-ranging gear, consolidations of cannery operations, fewer fish still, fewer and shorter runs yet, too much gear, and years of disastrous run failures.

There are some signs that modern mariculture and stream improvement sciences will bring back the teeming hordes of silver fish, but today there are more memories than annual summer adventures. Still, for many (only a century has really gone by since the day of the first canneries) the memories linger on—of the wonderful smell of Stockholm tar and oakum, of stale fish holds—the sound of windlasses

COURTESY OF WALTER YONKER

creaking in the dawn light as men strained their backs to the old hauling gear of the purse seines—the sounds of pings and pongs as the Chinamen went through the gleaming rows of stacked just-cooked salmon cooling in the warehouses, testing for leakers with a little hammer—the whistles of steamers coming in to tree-fringed inlets and wharves where canners had piles and piles of cased salmon ready to send to southland markets—tumultuous times, men stealing from each other in the trap pirate's day, men fighting with crowbars for favored beach fishing locations on Karluk Spit, men playing a funny kind of poker with officials in which the officials always seemed to win and the men got what they wanted in favored trap sites and seasons. Times when there were thirty-some thousand whites and thirty-some thousand Eskimos, Indians and Aleuts, and the Territory of Alaska was supported 75% by taxes from canned salmon. Good times and bad times, fortunes made and fortunes lost. The Big Gamble—hundreds of thousands of dollars in months-long preparation for The Big Pack—and more fish and more dollars than men ever dared dream of—or The Big Bust—fish that didn't show—streams that died and with them the salmon that would make or break another canner in the spawning-cycle years ahead.

Some called it a rape. Some called it the beginning of Alaska's permanency after the fleeting days of '98 and the riches of the gold fields had become sad memory.

Tremendous times to remember. When Ketchikan was "The Salmon Capital of the World" and 14 canneries ran full blast 2 or 3 months almost nonstop every summer and some longer. When girls walked the board sidewalks of Ketchikan, Wrangell, Craig, Cordova and dozens of other salmon canning communities, each lass in her knee-length rubber boots with the red tops, heading to the canneries for another shift in the endless summer toil of welcome wages and gleaming cans of red-fleshed salmon.

Remember the Filipino houses, the Indian houses, the all-alike little shacks strung out along the pile-supported boardwalks beside the canneries; the great beautiful billowing sails filling out on the square-riggers standing into the southwest winds off the Golden Gate, bending to a course for Bristol Bay; Italians and Swedes swabbing down, stowing shore gear, and settling in for the long haul North with a last jug, a crib board, and soiled decks of pinochle and pan and poker; the gulls wheeling; the ship bows rising, falling, rising again on the long, slow rollers; the tang of salt and tar and men, and on the horizon the promise of the season. Boom? Or Bust?

For those who remember, you will remember much more than is revealed between the photos and the ads dredged up for you in the following pages by editor and long-time Alaskan canneryman Laurence Freeburn. For those who did not know these times now past, all we can say is "This was a time. Those were days we will never forget. You should have been there."

Finally, our research revealed to us a tragedy in that so many photos of such tremendous events and people and times seem to have disappeared. Lest we be guilty of letting the last of the salmon canning memories also die, we have preserved that which we could find. Perhaps there are other great pictures, great stories, great ads we should have for some future publishing of another volume. We'll welcome receiving and hearing of what we missed.

For now, another pat on the back for salmon man Freeburn, your editor. And to our director of design, Roselyn Pape, who began this project as a sometimes weekend kayaker and now tells us she'd like to have sailed one of those Bristol Bay gill-netters. The salmon business is that way. It gets to you.

Red salmon, 1# talls, 48 to the case, 1919, $3.50 a case.
Red salmon, 1# talls, 48 to the case, 1976—$100?

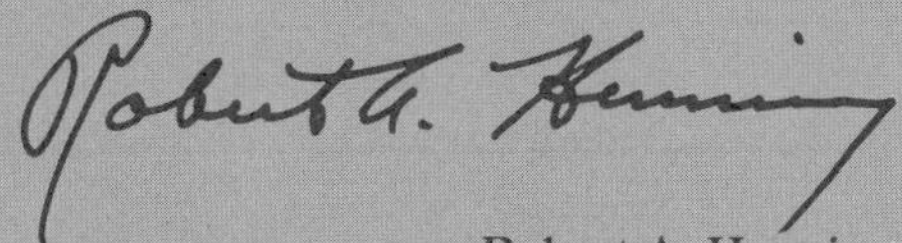

Robert A. Henning
President and Publisher

Left—*The apron brail comes up and suddenly the salmon are out of their element. Flying fish soon to quiet down and be canned for the market.*

The early morning visitor experienced the Indian village with smoke from the fish-drying houses, small gas boats and sailboats on the beach, a lone seine purse skiff, and the busy cannery with a schooner waiting for the pack. (Location unknown, can a reader identify?)

COURTESY OF WALTER YONKER

PREFACE

The photos on the following pages represent a lot of browsing in a lot of albums and record files. There is no real intent to try to accomplish an accurate chronology of the salmon industry in Alaska—great gaps of knowledge are apparent where no photos could be found—and many photos which seemed to be worthy of inclusion simply had no captions and in too many cases not even dates. It is unfortunate that those of us who have seen so many fascinating years go by in our industry did not keep better records, but perhaps this admittedly random sampling of old records and memories will encourage other cannerymen and fishermen to submit additional photos and information for another album later. As disconnected as the following material may be, we found ourselves fascinated to find things we had never known and to recall events and places and times too soon forgotten. Perhaps the readers of today will learn a bit of an important Alaska heritage—the days of the salmon men—and for those who were salmon men we are sure they will enjoy these bits and pieces as we have.

The history of the Alaska salmon canning industry, though studded with tales of adventure, glamour, and yes, even heroics, has nonetheless been in the final essence a continuing tale of corporate mortality.

There were fortunes made (most of them lost in following seasons), there were big seasons, tremendous runs that can still be remembered, but the rotting old stub piling and buildings not yet reclaimed by Mother Nature in hundreds of Alaskan bays and inlets are mute testimony that here is an industry dying.

One thing has not changed—the pure risk of the business. A fisherman builds a boat for a season not yet arrived. A packer spends a fortune in labor, supplies, insurance—all the necessary preseason expenditures—long before he has even seen the first fish. And then the expected run may simply not appear.

But money has not been the only risk. Early conditions were especially harsh. Even getting to some Alaskan packing locations by sail or by steam presented difficulties, and often conducting business great distances from the nearest towns, far from hospitals, butcher shops and hardware stores, made many a fish packer perhaps prematurely gray.

At one time, more than 75% of the Territory of Alaska's taxes came from canned salmon. That figure is far down today, and logs and oil have taken up where the salmon has left off, but there is a hope still remaining among the salmon people that modern science, money, and a concerned people may be able to bring back some measure of the great fish runs we knew in years past.

Remembering many times and many men, this book is dedicated to those good men who may derive some nostalgic pleasure from perusing these pages—the gambling packer, the salt-water man, the machinist, the blacksmith, the rigger, the tenderman, and the fishermen of many nationalities—men who built Alaska's salmon industry to its one-time glory—and to those who continue in the game today, gambling anew that this year will be the big one, that the industry will not die and that the labors and the trials of so many in the past need not have been in vain.

Laurence Freeburn

A steam-powered tender ready to brail a standing, or pile, trap. The pot scow can be seen to the right of the shack. These pilings were pulled at season's end and stored for the winter in a protected bay.

COURTESY OF WALTER YONKER

Plants

From: Moser, Jefferson F. *The Salmon and Salmon Fisheries of Alaska.* Washington: Government Printing Office, 1899. pp. 109-111.

KLAWAK

Klawak is on the western side of Prince of Wales Island, near the head of an inlet of the same name, which is an arm of Bucarelli Bay. It is off the regular line of travel, but is on the mail route between Wrangell and Howkan, and once a month a small mail steamer calls at the post office here. The oldest cannery in Alaska is in operation at Klawak, and in connection with it a steam sawmill and store. During the summer the settlement has a population of 300, of which number 275 are Indians and the rest white and mixed. There are about 50 houses in the vicinity, and also a school maintained by the Government. The cannery is one of two that have always been operated by Indian labor, and it is this enterprise that has made the village. The Indians employed are Haidas from the south, and Henegas (Thlinget) from the north and vicinity. The village is practically deserted during the winter, except by the cannery watchman and the storekeeper, the Indians going to their winter villages.

A saltery was operated at the place now known as Klawak until 1878, when it was purchased, with all its rights, etc., for $5,000 by the North Pacific Trading and

continued

REPRINTED FROM MOSER, SALMON FISHERIES OF ALASKA, PLATE 44

Many of Alaska's early canneries began operations as salteries. The Klawak (Klawock) cannery, above, established in 1878, generally described as "the first in Alaska" had been previously a summer fishing station for several years. At one time a Charles Waldron put up salt salmon at Klawak and later a George Hamilton put up salt salmon here. The Russians referred to the village of Klawak at some other nearby location, but not where it is today. For some time, the saltery location was known as Hamilton's Fishery.

PROVINCIAL ARCHIVES, VICTORIA, B.C.

Above—*This photo, probably taken in 1885 or 1886, is of the old salmon saltery at Loring, north of Ketchikan. It was owned by the Alaska Salmon Packing and Fur Company.* Right—*An early saltery at Ketchikan, date of photo unknown.*

Packing Company, and the present cannery was built that year. This was the pioneer in the salmon-packing business in Alaska and now enjoys the distinction of being the oldest cannery in operation, and of having made a pack every year since it was started. It has a capacity of 400 cases per day, but it has been worked to 700 cases when occasion demanded. The plant differs somewhat from those recently installed, in that less machinery is used and the pack is largely made by hand. The cutting, filling, and capping are all done by hand, and the cookers are old-fashioned, with dome-shaped or cylindrical covers that are raised by tackle to admit the trays, and then lowered and clamped. The plant, however, is very complete; the buildings are large and well-aired, there is plenty of room, the wharves are in good condition, and everything is thoroughly clean. It is said that the pack made here is second to none in the market.

continued

JOHN N. COBB COLLECTION, UNIVERSITY OF WASHINGTON LIBRARY

Salt Fish Statistics

PACK OF SALTED SALMON IN ALASKA, 1868 TO 1913.

YEAR	SALMON		SALMON BELLIES		DRY-SALTED SALMON	
	Barrels	Value	Barrels	Value	Pounds	Value
1868	2,000	$16,000	....	....	....	....
1869	1,700	13,600	....	....	....	....
1870	1,800	14,400	....	....	....	....
1871	700	6,300	....	....	....	....
1872	1,000	9,000	....	....	....	....
1873	900	7,200	....	....	....	....
1874	1,400	11,200	....	....	....	....
1875	1,200	9,600	....	....	....	....
1876	1,800	14,400	....	....	....	....
1877	1,950	15,700	....	....	....	....
1878	2,100	16,800	....	....	....	....
1879	3,500	28,000	....	....	....	....
1880	3,700	29,600	300	$3,300	....	....
1881	1,760	15,840	....	....	....	....
1882	5,890	53,010	....	....	....	....
1883	7,251	65,259	....	....	....	....
1884	6,106	54,954	....	....	....	....
1885	3,230	29,070	....	....	....	....
1886	4,861	43,749	....	....	....	....
1887	3,978	35,802	....	....	....	....
1888	9,500	85,500	....	....	....	....
1889	6,457	58,013	....	....	....	....
1890	18,039	162,351	....	....	....	....
1891	8,913	71,304	....	....	....	....
1892	17,374	140,057	53	$815	....	....
1893	24,005	120,083	....	....	....	....
1894	32,011	176,060	....	....	....	....
1895	14,234	85,404	....	....	....	....
1896	9,314	65,198	150	1,200	....	....
1897	15,848	110,936	2,846	28,460	....	....
1898	22,670	181,360	580	5,800	....	....
1899	22,382	167,865	235	2,350	....	....
1900	31,852	238,890	2,353	23,530	511,400	$10,288
1901	24,477	171,339	652	3,816	....	....
1902	30,384	212,688	328	2,952	....	....
1903	27,921	223,368	3,667	32,973	200,000	5,500
1904	13,674	89,209	208	1,950	966,812	16,180
1905	19,071	143,811	1,360	11,355	7,280,234	115,643
1906	17,283	126,194	1,338	13,644	1,107,680	16,969
1907	22,307	203,127	2,965	37,422	107,580	1,505
1908	31,472	266,713	7,600	85,994	20,800	416
1909	28,443	183,400	1,970	25,358	71,600	1,038
1910	12,779	111,634	1,626	19,007	22,178	554
1911	8,317	101,726	1,337	15,561	....	....
1912	34,602	305,928	37	606	....	....
1913	37,881	272,726	451	6,523	121,282	1,235
Total	**598,036**	**$4,564,368**	**30,056**	**$322,616**	**10,509,566**	**$169,268**

Source: *Pacific Fisherman,* January 1914, Vol. XII, No. 1, p. 82.

CLYDA SCHOTT GREELY COLLECTION. ALASKA HISTORICAL LIBRARY

Cannery at Wrangell, Alaska, mouth of Stikine River near the garnet ledge. Aberdeen Packing Company, with headquarters in Astoria, Oregon, sent Louis Schott and Bob Bell to Wrangell to operate this cannery. Schott is at left in the photo with a straw hat, Bell near the center with Indian paddle. The original photo said this was taken in 1887. Later documents declared this cannery was not built until 1889.

MRS. PETER NELSON COLLECTION. SAN FRANCISCO MARITIME MUSEUM

COURTESY OF HAROLD BRINDLE

Top—*An early salthouse at Koggiung, Bristol Bay.* Bottom—*For many years, into the thirties, many of the Native settlements along the Alaska Peninsula and in Bristol Bay included the picturesque old* barabaras, *half underground with a driftwood shell supporting an insulating roof and walls of sod.*

The cannery at Klawak is one of the smaller canneries of Alaska. It packs from 12,000 to 16,000 cases, the fish being obtained from the streams north and south of the cannery. The field was clear until 1896, when the cannery at Hunter Bay was built, and since that time the fishermen of the two canneries resort to the same streams on the south end of Prince of Wales Island. Ten years ago the Klawak cannery received fish from as far south as Hessa, and even from Nichols Bay, but in late years the principal sources of supply of redfish have been Hetta, the home stream, and Sar-Kar, with small intervening streams that some years have furnished from 1,000 to 2,000 redfish. Cohoes are nowhere very abundant in this district during the canning season, though the Indians say that large schools frequently come in after the cannery people leave. The Indian's idea of a large number, however, is usually measured by his own wants. He has no conception of a large number with reference to a cannery supply.

The Hetta stream has already been described.

The steam sawmill connected with the cannery has a capacity of 15,000 feet per day, but is only used for making boxes and supplying local demands.

The work at Klawak, from the taking of the fish in the streams to the loading of the cases containing the canned product on board the ocean carriers, is nearly all done by Indians—men, women, and children. From the earliest operations of the cannery until 1896 Indian labor has been exclusively employed, but in that year two Chinese were engaged, one for cap-cutting and the other for final testing and as general expert. In 1897 seven Chinese

continued

PHOTOS COURTESY OF HAROLD BRINDLE

Top—*Early cannery at Naknek.* Bottom—*Barrels of salt salmon being prepared for shipment. Photo was taken in 1922.* Right—*Splitting salmon for salting at Naknek Packing Company, 1926. Salting salmon preceded by many years the canning process.*

COURTESY OF HAROLD BRINDLE

Naknek Packing Company, 1924. This photograph, from the album of a canneryman, had this notation: "Ready for the start. What will the finish be?"

were employed, and the manager had under consideration the feasibility of using Chinese and doing away with Indian labor, chiefly upon the ground of economy, but also because Chinese labor is more certain and more easily controlled. If, at the beginning of or during a season, the cannery should decline to accede to the demands of the Indians for increased wages, a strike is apt to result, causing the loss of a pack.

In 1896 the cannery employed 2 white and 40 native fishermen; and in the cannery the help consisted of 7 whites, 32 klootchmen, 30 men and boys, and 2 Chinese. There were used 4 seines, each 110 fathoms long, and valued at $300 each. The steamer *Klawack*, of 11 tons, with a crew of 4, and valued at $5,000, and the steamer *Cora*, with a crew of 3, and valued at $1,000, were the vessels employed; but in spring and fall a vessel was chartered as a transport. One fish-scow valued at $100, and 4 seine-boats valued at $90 each, were also used.

Indian fishing huts on Skowl Arm, Kasaan Bay, 1898.

REPRINTED FROM MOSER, SALMON FISHERIES OF ALASKA, PLATE 36

The following table gives the packs for 1896 and 1897:

Species.	Dates.	Number of cases packed.	Number of fish per case.	Remarks.
	1896.			
Redfish	July 19 to Aug. 23	14,089	13	40 barrels of salmon bellies were salted, 300 cases of clams were packed (24 two-pound cans to the case), also 300 cases of clam juice (24 one-pint cans per case).
Cohoes	Aug. 15 to Sept. 25	2,073	8 to 9	
Humpbacks	July 24 to Aug. 10	513	22	
	1897.			
Redfish	June 26 to Aug. 31	9,520	13	200 cases of clams and 200 cases of clam juice were also packed.
Cohoes	Aug. 18 to Sept. 20	1,995	8	
Humpbacks	July 26 to Aug. 22	4,190	22	

continued

REPRINTED FROM MOSER, SALMON FISHERIES OF ALASKA, PLATE 52

Left—*Saltery at head of Uganuk Bay, Kadiak Island.*

Below left—*Saltery at Ideal Cove, Mitkof Island, 1908.* Below—*Deweyville, 35 miles north of Craig, Alaska, about 1907. Saltery of Fred Brockman, center.*

JOHN N. COBB COLLECTION, UNIVERSITY OF WASHINGTON LIBRARY

JOHN N. COBB COLLECTION, UNIVERSITY OF WASHINGTON LIBRARY

CLYDA SCHOTT GREELY COLLECTION, ALASKA HISTORICAL LIBRARY

MARTIN ASTRUP COLLECTION, SAN FRANCISCO MARITIME MUSEUM

Top—*The Lisianski Packing Company plant at Stag Bay, Lisianski Strait, with the survey steamer* Patterson *at the wharf, early 1900's. The building at right had a mess house downstairs, superintendent's quarters and guest rooms upstairs.* Bottom—*Sledging to the Coffee Creek cannery, Bristol Bay, early in the season. Photo taken about 1920.*

The sawmill and its machinery are valued at $13,000. The machinery also furnishes the motive power for the cannery. The cannery buildings, complete with storehouse, wharves, tramways, machinery, tools, fixtures, retorts, etc., are valued at a little under $20,000. The whole property, including all the above, cannery and sawmill, buildings, dwellings, wharves, tools, fixtures, and machinery, steamers, boats, seines, etc., but exclusive of material on hand, is valued at about $50,000.

When salmon were being packed it was noticed that the butchers on the fish float threw all the heads into canoes waiting alongside to receive them, and many canoe loads were carried away. Upon inquiry it was learned that the heads are put in baskets or bags placed along the shores between the high and low water marks, weighted with stones to keep them in place, and left for a week or ten days until thoroughly ripe; the snouth or nose is then cut off and consumed by the Indians as a great delicacy.

—Moser, *Salmon Fisheries of Alaska,* pp. 109-111.

WILLIAM L. OWENS COLLECTION, SAN FRANCISCO MARITIME MUSEUM

Left—*The* A. J. Fuller *at Northwest Fisheries Company cannery, Uyak, Kodiak Island, Alaska, in early 1920's.*

W. H. MEYER COLLECTION, SAN FRANCISCO MARITIME MUSEUM

COURTESY OF JOHN GILBERT

Left—*Handling fish from the lighter the hard way—1912.*
Above—*Discharging fish at Egegik into car on rails, about 1908. This very primitive method of getting fish ashore was later replaced by the mechanical elevator.*

From: Moser, *Salmon Fisheries of Alaska*, pp. 49-50.

For four years after salmon canning was begun in Alaska it was confined to the southeastern part of the Territory; in 1882 the business was inaugurated in the Cook Inlet and the Kadiak regions; two years later canning commenced in Bering Sea, and in 1889 in Prince William Sound. In all of these sections the industry has continued and has undergone a great increase, so that the pack of 1897 was larger than that of any previous year except 1896. Up to and including 1897 more than two-fifths of the canned products, namely, 3,286,505 cases, had come from the Kadiak and Chignik district; somewhat more than one-fifth, or 1,739,832 cases, from southeast Alaska, and a little less than one-fifth, or 1,496,513 cases, from Bering Sea.

The number of salmon canneries in operation in 1897 was 29, divided as follows among the different districts: Southeast Alaska, 9; Prince William Sound and Copper River, 2; Cook Inlet, 1; Kadiak and Chignik, 10, and Bering Sea, 7. In the previous year, also, 29 canneries were operated, 9 being in southeast Alaska, 3 on Prince William Sound and Copper River, 1 on Cook Inlet, 8 at Kadiak and Chignik, and 8 in Bering Sea.

The number of persons employed in the Alaskan salmon-canning business in 1897 was 5,252. Of these, 1,148 were white fishermen, 759 native fishermen, 312 white cannery employees, 439 native cannery hands, 2,268 Chinese cannery hands, and 326 hatchery men, crews of cannery launches and other vessels, and various other employees. The persons employed in various capacities in each district numbered as follows: Southeast Alaska, 1,829; Prince William Sound and Copper River, 364; Cook Inlet, 173; Kadiak and Chignik, 1,577, and Bering Sea, 1,309.

The amount of invested capital represented by the Alaskan salmon industry is very large. The value of the vessels, boats, fishing gear, buildings, machinery, etc., including the canneries not operated but held in reserve, aggregated approximately $3,623,200 in 1897, apportioned as follows among the different districts: Southeast Alaska, $597,400; Prince William Sound and Copper River, $263,500; Cook Inlet, $154,300; Kadiak and Chignik, $1,741,000, and Bering Sea, $867,000. By far the largest item in the investment was the cannery buildings and equipment, which had a value of $2,630,860. The other leading items were vessels and boats, $836,100; 544 gill nets, $69,470; 48 traps, $48,050, and 133 seines, $38,680.

Table showing the output of the salmon canneries of Alaska from 1878 to 1897.

Name of company and location of cannery.	1878.	1879.	1880.	1881.	1882.	1883.	1884.
SOUTHEAST ALASKA.	*Cases.*	*Cases.*	*Cases.*	*Cases.*	*Cases.*	*Cases.*	*Cases.*
Cape Fox Packing Co., north shore Boca de Quadra						Packs unknown.	
North Pacific Trading and Packing Co., Klawak, Prince of Wales Island	5,402	6,675	6,539	8,977	11,501	8,240	6,189
Cutting Packing Co., Old Sitka, Baranof Island	2,757	5,855	Closed.	Closed.	(1)		
Chilkat Packing Co., east shore Chilkat Inlet						6,000	7,000
Northwest Trading Co., Pyramid Harbor, Chilkat Inlet						3,800	6,000
Totals of annual packs in southeast Alaska	8,159	12,530	6,539	8,977	11,501	[2] 18,040	[2] 19,189
COOK INLET.							
Alaska Packing Co., Kussilof River, Cook Inlet					6,044	14,818	21,141
KADIAK AND CHIGNIK.							
Karluk Packing Co., Karluk, Kadiak Island					4,200	13,479	20,156
BERING SEA.							
Arctic Packing Co., Nushagak River, Bristol Bay							[3] 400

Name of company and location of cannery.	1885.	1886.	1887.	1888.	1889.	1890.	1891.
SOUTHEAST ALASKA.	*Cases.*	*Cases.*	*Cases.*	*Cases.*	*Cases.*	*Cases.*	*Cases.*
Cape Fox Packing Co., north shore Boca de Quadra	Packs unknown.		New	company;	moved to	Ketchikan in 1887.	
Tongass Packing Co., Ketchikan, Tongass Narrows			5,500	11,000	13,000	Burnt, Aug., 1889.	
Metlakahtla Industrial Co., Metlakahtla, Annette Island						[3] 500	7,000
Alaska Salmon Packing and Fur Co., Loring, Naha Bay				18,771	28,862	23,024	22,786
Boston Fishing and Trading Co., Yes or McDonald Bay					4,500	9,343	17,365
Cape Lees Packing Co., Burroughs Bay, Behm Canal				5,732	10,574	10,823	
Aberdeen Packing Co., Stikine River			3,400	14,000	Moved;	new company.	
Glacier Packing Co., Point Highfield, Wrangell Island					13,800	14,600	15,876
North Pacific Trading and Packing Co., Klawak, Prince of Wales Island	8,428	7,860	9,562	12,325	11,370	10,188	9,256
Baranoff Packing Co., Redoubt, Baranof Island					4,454	10,123	Moved.
Baranoff Packing Co., Redfish Bay, Baranof Island							7,949
Astoria and Alaska Packing Co., Freshwater Bay, Chichagof Island					Unknown.	Moved.	
Astoria and Alaska Packing Co., Point Ellis, Kuiu Island						8,000	16,200
Bartlett Bay Packing Co., Bartlett Bay, Icy Straits					4,300	12,000	7,600
Chilkat Packing Co., east shore Chilkat Inlet	2,400	1,700	8,000	Closed.	13,500	15,000	13,369
Chilkat Canning Co., Chilkat Village, Chilkat Inlet					19,000	17,000	20,914
Northwest Trading Co., Pyramid Harbor, Chilkat Inlet	Closed.	8,600	5,000	New Co.			
Pyramid Harbor Packing Co., Pyramid Harbor, Chilkat Inlet				19,300	[4] 13,400	12,300	18,300
Totals of annual packs in southeast Alaska	[2] 10,828	[2] 18,160	31,462	81,128	[2] 136,760	142,901	156,615
PRINCE WILLIAM SOUND AND COPPER RIVER.							
Central Alaska Co., Little Kayak Island, Gulf of Alaska					1,600	Moved to Thin Point	
Peninsula Trading and Fishing Co., Little Kayak Island, Gulf of Alaska					2,540	14,200	Moved.
Peninsula Trading and Fishing Co., Coquenhena, Copper River Delta							18,085
Pacific Packing Co., Odiak, Prince William Sound					5,064	13,716	27,209
Pacific Steam Whaling Co., Odiak, Prince William Sound					15,000	14,278	22,797
Totals of annual packs in Prince William Sound and Copper River					24,204	42,194	68,091
COOK INLET.							
Northern Packing Co., Kenai				12,996	18,712	15,905	18,254
George W. Hume, Kussilof River						12,750	21,586
Alaska Packing Co., Kussilof River	NewCo.						
Arctic Fishing Co., Kussilof River	19,217	28,433	30,765	29,455	31,782	(5)	19,157
Totals of annual packs in Cook Inlet	19,217	28,433	30,765	42,451	50,494	28,655	58,997
KADIAK AND CHIGNIK.							
Kodiak Packing Co., Karluk, Kadiak Island				26,146	30,287	43,060	41,000
Hume Packing Co., Karluk, Kadiak Island					28,000	37,613	36,247
Aleutian Islands Fishing and Mining Co., Karluk, Kadiak Island				33,700	53,551	40,335	[6] 39,312
Karluk Packing Co., Karluk, Kadiak Island	33,470	46,150	71,750	101,304	62,057	40,306	66,483
Alaska Improvement Co., Karluk, Kadiak Island					25,600	26,000	26,000
Arctic Packing Co., Alitak Bay, Kadiak Island					13,850	17,800	26,600
Kodiak Packing Co., Alitak Bay, Kadiak Island					12,535	16,347	(7)
Arctic Packing Co., Larsen Cove, Uyak Bay				37,500	44,200	37,600	[8] 41,000
Royal Packing Co., Afognak Island					16,412	9,926	(9)
Russian-American Packing Co., Afognak Island					25,500	26,500	[10] 25,000
Chignik Bay Co., Chignik Bay, Alaska Peninsula					21,500	14,455	24,730
Shumagin Packing Co., Chignik Bay, Alaska Peninsula					11,200	[11] 14,455	[11] 24,832
Chignik Bay Packing Co., Chignik Bay, Alaska Peninsula					10,500	[11] 14,455	[11] 24,780
Western Alaska Packing Co., Ozernoi, Stepovak Bay, Alaska Peninsula					6,400	2,198	Aband'd.
Thin Point Packing Co., Thin Point, Alaska Peninsula					25,161	2,401	4,206
Central Alaska Co., Thin Point, Alaska Peninsula						7,000	4,089
Totals of annual packs in Kadiak and Chignik	33,470	46,150	71,750	198,650	386,753	350,451	384,279

continued

Table showing the output of the salmon canneries of Alaska from 1878 to 1897—Continued.

Name of company and location of cannery.	1878.	1879.	1880.	1881.	1882.	1883.	1884.
BERING SEA.							
Arctic Packing Co., Nushagak River, Bristol Bay	14,000	19,000	24,000	25,000	25,000	33,000	30,883
Alaska Packing Co., Nushagak River, Bristol Bay		16,500	27,500	19,000	30,000	31,000	31,077
Bristol Bay Canning Co., Nushagak River, Bristol Bay		13,322	21,200	30,000	33,221	30,400	37,100
Nushagak Packing Co., Nushagak River, Bristol Bay				15,886	27,764	23,990	30,363
Bering Sea Packing Co., Ugashik River, Bristol Bay							3,995
Totals of annual packs in Bering Sea	14,000	48,822	72,700	89,886	115,985	118,390	133,418

Name of company and location of cannery.	1892.	1893.	1894.	1895.	1896.	1897.
SOUTHEAST ALASKA.	*Cases.*	*Cases.*	*Cases.*	*Cases.*	*Cases.*	*Cases.*
Quadra Packing Co., Mink Bay, Boca de Quadra					8,000	24,500
Metlakahtla Industrial Co., Metlakahtla, Annette Island	11,125	12,500	14,000	12,000	17,650	15,490
Pacific Steam Whaling Co., Hunter Bay, Prince of Wales Island					27,442	34,388
Alaska Salmon Packing and Fur Co., Loring, Naha Bay	21,446	25,153	26,869	32,554	61,467	62,040
Boston Fishing and Trading Co., Yes or McDonald Bay	13,734	15,162	12,000	14,100	24,100	22,300
Cape Lees Packing Co., Burroughs Bay, Behm Canal	Closed.	Closed.	Dismantled.			
Glacier Packing Co., Point Highfield, Wrangell Island	Closed.	22,728	25,250	27,416	44,233	45,918
North Pacific Trading and Packing Co., Klawak, Prince of Wales Island	10,194	12,595	14,455	12,228	16,675	15,705
Baranoff Packing Co., Redfish Bay, Baranof Island	10,259	9,889	11,189	14,805	15,358	14,070
Astoria and Alaska Packing Co., Point Ellis, Kuiu Island	Burnt.[1]					
Bartlett Bay Packing Co., Bartlett Bay, Icy Straits	Closed.	Closed.	Dismantled.			
Chilkat Packing Co., east shore Chilkat Inlet	Burnt.					
Chilkat Canning Co., Chilkat Village, Chilkat Inlet	20,000	24,418	Closed.	Closed.	Closed.	Closed.
Pyramid Harbor Packing Co., Pyramid Harbor, Chilkat Inlet	28,964	13,668	38,781	35,373	47,456	37,456
Totals of annual packs in southeast Alaska	115,722	136,053	142,544	148,476	262,381	271,867
PRINCE WILLIAM SOUND AND COPPER RIVER.						
Peninsula Trading and Fishing Co., Coquenhena, Copper River Delta	Closed.	15,270	15,000	15,000	20,672	Closed.
Pacific Packing Co., Odiak, Prince William Sound	Closed.	28,999	28,378	21,453	39,873	23,301
Pacific Steam Whaling Co., Odiak, Prince William Sound	Closed.	32,729	35,285	Moved.		
Pacific Steam Whaling Co., Orca, Prince William Sound				23,041	32,321	28,756
Totals of annual packs in Prince William Sound and Copper River		76,998	78,663	59,494	92,866	52,057
COOK INLET.						
Northern Packing Co., Kenai	Closed.	Closed.	Closed.	Closed.	Closed.	Closed.
Pacific Steam Whaling Co., Kenai						(2)
George W. Hume, Kussilof River	20,741	Closed.	(3)			
Arctic Fishing Co., Kussilof River	Closed.	31,665	34,033	36,188	34,767	32,532
Totals of annual packs in Cook Inlet	20,741	31,665	34,033	36,188	34,767	32,532
KADIAK AND CHIGNIK.						
Kodiak Packing Co., Karluk, Kadiak Island	Closed.	30,138	Closed.	Closed.	Closed.	Closed.
Aleutian Islands Fishing and Mining Co., Karluk, Kadiak Island	Consol.					
Hume Aleutian Packing Co., Karluk, Kadiak Island	76,233	59,959	79,000	47,500	70,320	49,633
Karluk Packing Co., Karluk, Kadiak Island	75,416	59,220	79,000	48,379	68,495	54,777
Alaska Improvement Co., Karluk, Kadiak Island	52,098	43,076	54,300	35,700	87,613	49,852
Hume Canning and Trading Co., Tanglefoot Bay, near Karluk		15,429	26,984	15,277	Closed.	Closed.
Arctic Packing Co., Alitak Bay, Kadiak Island	21,077	25,777	27,720	15,331	23,155	37,401
Kodiak Packing Co., Alitak Bay, Kadiak Island		Mach. moved.				
Pacific Steam Whaling Co., Uyak Anchorage, Kadiak Island						17,000
Hume Bros. & Hume, Uyak Anchorage, Kadiak Island						13,375
Arctic Packing Co., Larsen Cove, Uyak Bay	Closed.	Closed.	Closed.	Closed.	Mach. moved.	
Uganuk Fishing Station, Uganuk Bay, Kadiak Island					21,005	2,113
Royal Packing Co., Afognak Island	Closed; inoperative by President's proclamation creating a Fish Commission reservation.				Mach. moved.	
Russian-American Packing Co., Afognak Island					Disman.	
Chignik Bay Co., Chignik Bay, Alaska Peninsula	49,931	57,553	55,352	70,050	48,361	38,159
Shumagin Packing Co., Chignik Bay, Alaska Peninsula	Consol.					
Chignik Bay Packing Co., Chignik Bay, Alaska Peninsula	Closed.	Closed.	Closed.	Closed.	Closed.	Closed.
Hume Bros. & Hume, Chignik Bay, Alaska Peninsula					17,893	12,000
Pacific Steam Whaling Co., Chignik Bay, Alaska Peninsula					21,515	24,000
Thin Point Packing Co., Thin Point, Alaska Peninsula	Closed.	Closed.	Disman.			
Central Alaska Co., Thin Point, Alaska Peninsula	Closed.	Closed.	Closed.	Disman.		
Totals of annual packs in Kadiak and Chignik	274,755	291,152	322,356	232,237	358,357	298,310
BERING SEA.						
Arctic Packing Co., Nushagak River, Bristol Bay	Closed.	35,848	30,413	33,631	35,676	35,890
Alaska Packing Co., Nushagak River, Bristol Bay	31,859	37,188	30,038	34,632	39,115	37,849
Bristol Bay Canning Co., Nushagak River, Bristol Bay	31,640	34,750	30,990	33,434	38,314	34,117
Nushagak Packing Co., Nushagak River, Bristol Bay	Closed.	Closed.	Closed.	Closed.	Closed.	Closed.
Point Roberts Packing Co., Kvichak River, Bristol Bay					29,730	55,508
Arctic Packing Co., Naknek River, Bristol Bay				22,731	27,133	34,676
Naknek Packing Co., Naknek River, Bristol Bay				13,700	8,600	18,000
Bering Sea Packing Co., Ugashik River, Bristol Bay	Closed.	Closed.	17,394	12,007	20,004	Closed.
Ugashik Fishing Station, Ugashik River, Bristol Bay					19,764	38,272
Totals of annual packs in Bering Sea	63,499	107,786	108,844	150,135	218,336	254,312

[1] Burnt May 1. [2] Buildings erected. [3] Consolidated with the Arctic Fishing Company.

[1] Machinery moved; site abandoned.
[2] The records for the southeast Alaska district for 1883, 1884, 1885, 1886, and 1889 lack the pack of one small cannery each, the output of which can not be ascertained.
[3] Experimental pack.
[4] Burnt and rebuilt in spring.
[5] Vessel lost; closed.
[6] Packed in cannery of Hume Packing Co. at Karluk.
[7] Consolidated with the Arctic Packing Company.
[8] Packed in cannery of Kodiak Packing Company at Karluk.
[9] The quota of fish for the Royal Packing Company at Afognak was packed by the Karluk Packing Company at Karluk.
[10] Packed in cannery of Alaska Improvement Company at Karluk.
[11] Packed in cannery of Chignik Bay Company.

Source: Moser, *Salmon Fisheries of Alaska*, pp. 51-52.

COURTESY OF WALTER YONKER

Above—*Mechanical elevators were loaded by hand in 1910.* Right—*Two unknown cannery men in what appears to be a cannery office.*

COURTESY OF HAROLD BRINDLE

From: Moser, *Salmon Fisheries of Alaska,* p. 17.

In 1888 the following additional canneries were in operation: Alaska Salmon Packing and Fur Company, at Loring, southeast Alaska; Cape Lees Packing Company, at Burroughs Bay, southeast Alaska; Northern Packing Company, at Kenai, Cook Inlet; Kodiak Packing Company and Aleutian Islands Fishing and Mining Company, both at Karluk, Kadiak Island; Arctic Packing Company, at Larsen Bay, Kadiak Island, and the Nushagak Packing Company, on the Nushagak River, Bering Sea.

The increase of canneries in 1888 and their large output called the attention of the public to the Alaska salmon fisheries, and in 1889 there was a further increase of canneries, as follows:

Southeast Alaska: Boston Fishing and Trading Company, Yes (McDonald) Bay; Baranoff Packing Company, at the Redoubt, near Sitka; Astoria and Alaska Packing Company, Freshwater Bay; Bartlett Bay Packing Company, Bartlett Bay, and Chilkat Canning Company, Chilkat Village.

Prince William Sound and Copper River: Peninsula Trading and Fishing Company, Little Kayak Island; Central Alaska Company, Little Kayak Island; Pacific Packing Company, Eyak, Copper River, and the Pacific Steam Whaling Company, Eyak, Copper River.

Kadiak and Chignik: Hume Packing Company, Karluk; Alaska Improvement Company, Karluk; Arctic Packing Company, Alitak Bay; Kodiak Packing Company, Alitak Bay; Royal Packing Company, Afognak Island; Russian-American Packing Company, Afognak Island; Chignik Bay Company, Chignik Bay; Shumagin Packing Company, Chignik Bay; Chignik Bay Packing Company, Chignik Bay; Western Alaska Company, Ozernoi, Alaska Peninsula; Thin Point Packing Company, Thin Point, Alaska Peninsula.

continued

The cannery of the Alaska Improvement Company, at Karluk, was built and ready for operation in 1888, but the loss of the *Julia Ford*, the cannery ship, with all the season's outfit on board, kept the cannery closed, and the first pack was therefore made in 1889.

In 1888 there were 17 canneries in operation in Alaska with an output of 412,115 cases, and in 1889 there were 37 canneries with an output of 714,196 cases. Two more canneries were added to the list of 1890—George W. Hume, at Kussilof, Cook Inlet, and the cannery of the Metlakahtla Industrial Company, in southeast Alaska. In 1891 only one new cannery, that of the Bering Sea Packing Company, at Ugashik, Bering Sea, was added, while several operated the year previous were closed, and in a few other cases there was a consolidation of interests. In 1892 and 1893 there was a further consolidation of cannery interests, and in the latter year one additional cannery was operated, that of the Hume Canning and Trading Company, in Tanglefoot Bay, near Karluk. There were no new canneries built in 1894, but in 1895 two more were in operation at Naknek, in Bering Sea, the Arctic Packing Company and the Naknek Packing Company.

In 1896 the following new companies were in operation:

Southeast Alaska: Quandra Packing Company, in Mink Arm, Boca de Quandra; Pacific Steam Whaling Company, Hunter Bay.

Kadiak Island: Uganuk Fishing Station, Uganuk Bay.

Alaska Peninsula: Hume Brothers & Hume, Chignik Bay; Pacific Steam Whaling Company, Chignik Bay.

Bering Sea: Point Roberts Packing Company, at Koggiung, Kvichak River; Ugashik Fishing Station, Ugashik River.

—Moser, *Salmon Fisheries of Alaska,* p. 17.

SAN FRANCISCO MARITIME MUSEUM

Discharging 40,000 salmon from lighter at a cannery, Bristol Bay, Alaska, via mechanical elevator, circa 1920.

SAN FRANCISCO MARITIME MUSEUM

Left—*Waiting for their turn to unload—1959.*

AXEL WINDERSTROM COLLECTION, SAN FRANCISCO MARITIME MUSEUM

Above—*Discharging fish, then to anchor and sleep.* Right—*Bristol Bay boats and scow.*

HARRY JOHANNESEN COLLECTION, SAN FRANCISCO MARITIME MUSEUM

Successful Performance of Self-Unloading Fish Scow

From: *Pacific Fisherman*, February 1917, p. 20.

NEW MODEL FISH SCOW.

One of the most interesting new pieces of salmon cannery equipment which is being shown this season, is the Cook & Darrell self unloading scow.

Simplicity of operation, extreme low cost of installation, valuable labor and time saving, are a few of the many claims made for this system. Entire elimination of the unsanitary and undesirable pughing and mutilation of the fish perhaps predominates above all other features, and has proven the leading point in its favor. That the packer has long been making this feature of his business a thorough study, with the full knowledge that in the near future some new system would be adopted, is evidenced by the deep interest taken in the Cook and Darrell system. In discussing this, W. H. Gramm, Cook and Darrell's agent, says:

"There is no difference in brailing into the scow with this system than with the scows now in use excepting the added feature that instead of having to work the fish into the separate bins, when there is a small load, the system of raised floors will keep the fish distributed properly so there will be no unusual list to the scow.

"When the scow arrives at the cannery with the load of fish, the chute is placed in position and regulated for the even flow of the fish to the conveyer. The gate or trap is then opened and the automatic unloading begins. With the use of a hose with a fair volume of water, the fish, through the arrangement of bins, floors, etc., will unload as fast as the conveyer will take them away. The bins are so arranged that the scow will at all times list toward the conveyer, the angle of the list desired being regulated automatically.

"In many instances it is found that the conveyer does not run low enough in the water but with a slight change and minimum cost this can be altered to suit the system.

From: *Pacific Fisherman*, October 1917, p. 35.

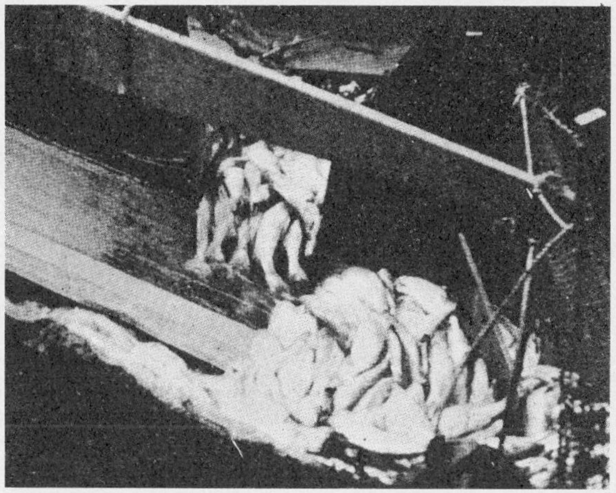

The Cook & Darrell self-unloading fish scow, introduced to the salmon packing industry this season for the first time, has been thoroughly tried out during the last few months at the plant of the Everett Packing company at Everett, Wash., and the results not only demonstrate the correctness of the mechanical principles involved, but show conclusively that the new type of scow will do all that is claimed for it under practical every-day working conditions.

In expressing his satisfaction with the scow, J. O. Morris, vice-president of the Everett company, says: "Owing to the arrangement for delivering the fish from the opposite side of the scow through a chute to the conveyor, causing the scow to list toward point of discharge, the fish were automatically delivered to the conveyor faster than it could take them away. This, however, is merely a matter of a bigger and more powerful conveyor, which can be regulated to unload the fish at any desired speed. The principle of the scow is right and means not only the saving of a lot of high-priced labor, but also the re-handling and bruising of the fish, which is no small item itself. Wherever possible, we will equip our fish scows with this device."

The accompanying illustrations show the actual working of the new scow as contrasted with the old way of unloading by the use of peughs and "elbow grease". The upper left-hand view shows the scow starting to unload, which is accomplished with practically no labor when once the chute is connected with the conveyor and the gate opened, only one man being required to regulate the flow and prevent swamping of the conveyor. The upper right-hand view shows the rush of fish to the elevator: while the lower picture vividly illustrates the automatic-unloading feature of the device.

Right—*Pacific Steam Whaling Company salmon cannery at Anchorage Bay, near Chignik, 1897.*

COURTESY OF HAROLD BRINDLE

REPRINTED FROM MOSER, SALMON FISHERIES OF ALASKA, PLATE 62

COURTESY OF HAROLD BRINDLE

Above—*Hoonah Packing Company, 1917.* Left—*Loring, near Ketchikan, 1920.*

REPRINTED FROM MOSER, SALMON FISHERIES OF ALASKA, PLATE 19

REPRINTED FROM MOSER SALMON FISHERIES OF ALASKA, PLATE 49

Views of oil and guano factory at Killisnoo, late 1890's.

From: Moser, *Salmon Fisheries of Alaska,* pp. 121-123.

KILLISNOO

Killisnoo is on the northeastern side of Kenasnow Island, close to the western shore of Admiralty Island, opposite Peril Straits, and about 3 miles south of the entrance to Kootznahoo Inlet. It is a regular calling-place for the steamers of the Pacific Coast Steamship Company. It has a post office, store, Government school, and a Greek chapel. In 1890 the regular population was 79. The works of the Alaska Oil and Guano Company, incorporated with a capital stock of $75,000, are located here. The business of the company is to extract the oil from the herring and convert the refuse into guano. Incidentally a few salmon and herring are salted.

In 1880 Messrs. Spuhn & Vanderbilt came to Alaska, operating under the title of the Northwest Trading Company, for the purpose of establishing stores for trade with the natives. One store was located at Killisnoo, an old Hudson Bay Company station.

As herring were plentiful in the inlet at Kootznahoo, a small plant was installed in 1882 for extracting oil, simply in an experimental way, and as it proved successful it was gradually enlarged until its present capacity is said to be 1,800 barrels of herring in twenty-four hours, or 350,000 gallons of oil, 1,500 tons of guano, and 1,000 half-barrels of salt herring for the season. As prices are now low, the works are not run to their full capacity. In 1884 the plant for the manufacture of guano was installed and operated. In 1888 the Northwest Trading Company, which also built the Pyramid Harbor salmon cannery in 1883, went into the hands of trustees, and the present company was organized.

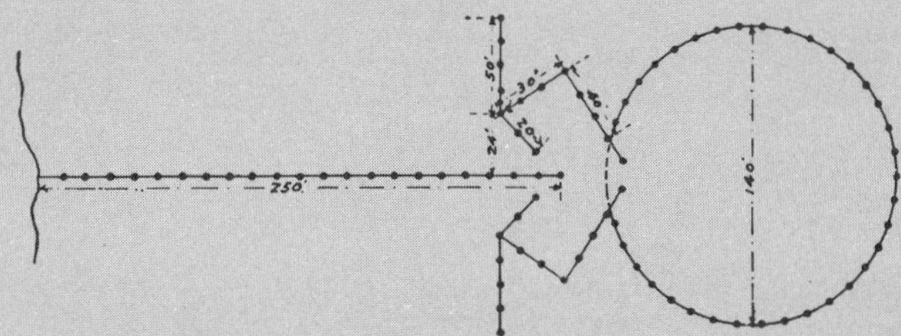

Sketch of herring trap at Kootznahoo Inlet.

The steamer *Dolphin,* of 60 tons net register, valued at $10,000, with a crew of 18, including fishermen, is used for cruising through interior waters for fish. The steamers *Favorite* and *Louise* operate in the inlet with

continued

scows, using 3 purse seines, 125 to 150 fathoms long, 12 fathoms deep, ½-inch stretched mesh, valued at $1,000 each, fully rigged. The *Favorite* is 42 tons net, and is valued at $5,000; it has a crew of 16, including fishermen, viz, 6 whites, 9 natives, and 1 Chinese cook. The steamer *Louise* is 5 tons net, valued at $3,000, and has a crew of 3. Three lighters, worth $800 each, and 8 seine boats, valued at $125 each, besides numerous skiffs, etc., are employed. This year a trap was driven in the inlet, but at the time of our visit it had not proved satisfactory. In and about the factory 17 whites, 35 natives, 3 Chinese, and 5 Japanese are employed.

The white fishermen are paid $50 a month, and board; the native fishermen get $1.50 a day, or practically $45 a month, for they are paid even if detained, and while board is not stipulated they practically get it. Laborers generally have $1 a day, but about 10 of them, who have been employed a long time and are faithful, having the more difficult work to do, receive $1.50 per day. Boys are paid 50 cents a day. A good native fisherman or laborer makes about $200 a season. All wages are paid in cash, the company disbursing $10,000 to $15,000 a year to the natives and the few white men who make Killisnoo their home.

The factory is ready for operation from July 1 to December 31.

In 1896 the output was as follows: From 27,750 barrels of herring there were made 90,650 gallons of oil and 550 tons of guano. This is the smallest output of oil since the second year of operation. 250 half-barrels of herring, 25 barrels of salmon, and 150 half-barrels of salmon bellies were salted.

—Moser, *Salmon Fisheries of Alaska,* pp. 121-123.

PHOTOS REPRINTED FROM MOSER, SALMON FISHERIES OF ALASKA, PLATE 14

Top—*Indian fish-trap and artificial channel in stream, first inlet, south side of Moira Sound, 1897.*
Bottom—*Barrier in stream entering first inlet on south side of Moira Sound, 1897.*

From: Moser, *Salmon Fisheries of Alaska,* pp. 66-67.

METLAKAHTLA

Metlakahtla, or New Metlakahtla, is situated on Port Chester, on the western side of Annette Island. It consists of a community of Tsimpsean Indians who, in 1887, followed Mr. William Duncan from Metlakahtla, British Columbia, to make a home in Alaska. So much has been written of this interesting colony that a history of their progress is superfluous in this report. Suffice it to say, that all honor is due to the energetic and intelligent civilizer and preacher, Mr. Duncan. . . . After having brought them to a civilized condition through his own example and energetic teachings, some question arose in the church, and he deemed it best to retire. He was followed by about 400 of the community, and, as they were not allowed to carry with them their own belongings, they calmly faced the privations they were to endure in their new home, and to-day the New Matlakahtla stands a monument to their skill and industry and an honor to the self-sacrificing devotion of their leader.

Annette Island has been set aside by an act of Congress for a reservation for the Metlakahtlans and other Alaskan natives who may join them, under regulations prescribed by the Secretary of the Interior. An attempt has been made by some influential whites to open mines on the island. In 1890 the population of Metlakahtla was 823, of which number 817 were Indians, 4 whites, and 2 mixed.

One of the principal industries of this community is salmon canning, which is carried on by an organization known as the Metlakahtla Industrial Company. When the colony moved to their new home in 1887, after the people were comfortably housed, Mr. Duncan commenced erecting buildings with a view to establishing a cannery. It took several years to obtain the necessary funds and install the machinery, and it was not until 1890 that any pack was attempted. In that year an experimental pack of only 500 cases was made. The following year it was in active operation and a pack of 5,834 cases was the result.

All parts and branches of the cannery and the fisheries supplying it are conducted by native Indians, who show

continued

REPRINTED FROM MOSER, SALMON FISHERIES OF ALASKA, PLATE 23

Salmon cannery at Metlakahtla in 1890's.

From: *Pacific Fisherman,* April 1917, p. 16.

METLAKAHTLA INDIAN FISHING RIGHTS UPHELD.

The U. S. Circuit Court of Appeals at San Francisco on March 19 handed down a decision upholding the fishing rights of the Metlakahtla Indians in the suit brought by the Government, on behalf of the natives, against the Alaska-Pacific Fisheries. This sustains a decision given out some time ago by Judge Robert W. Jennings in the U. S. District Court of Alaska, in which the fishing reserve made around Annette island by presidential proclamation was declared valid.

The order proclaimed a fishing reserve for a distance of 3,000 feet in the waters surrounding the island, for the exclusive use of the natives. The suit was started last year, when the Alaska-Pacific Fisheries built a fish trap within the district. Under the decision, the Annette Island Packing company, which operates under a five-year agreement with the Indians, will have the exclusive cannery privileges of the island during that time.

The following table shows the packs at Metlakahtla for 1896 and 1897:

Species.	1896.				1897.	
	No. of cases.	Number of fish per case.	Value per case.	Date of packing.	No. of cases.	Date of packing.
Redfish	8,280	*12	$3.80	July 1 to Sept. 5	7,090	July 8 to Sept. 2.
Cohoes	1,500	6 or 7	3.40	Aug. 1 to Sept. 10	840	July 20 to Sept. 2.
Humpbacks	7,420	21 or 22	2.40	July 21 to Aug. 22	7,260	July 20 to Aug. 26.
Do-overs	450	....	3.00	July 1 to Sept. 10	300	
Total	17,650	....	....	....	15,490	

* Quadra redfish ran 8 to the case; those from Moira Sound, 15 to the case. The average was about 12.

Source: Moser, *Salmon Fisheries of Alaska*, p. 67.

continued

COURTESY OF WALTER YONKER

considerable skill in the different kinds of work. During the time of our first visit, June 10 to 12, they were making cans for the season's pack, rebuilding the steam tender, and some 30 or 40 women and girls were making seines and nets for the use of the community and for sale. All this work is done by the Indians, and apparently well done.

The value of the cannery buildings, machinery, piping, tools, material, etc., is a little under $25,000. In 1896 the product of 72 native fishermen was used, and during August 162 natives were employed in the cannery. In the fisheries they used 9 drag seines, each 100 fathoms in length, valued at $200 each; and 2 purse seines, 120 fathoms long, worth $350 each. The vessels and boats employed were the steam tender *Marie G. Haaven* (rebuilt in the spring of 1897), valued at $3,000, of 12.8 net tons, and with a crew of 5; also 7 Columbia River boats, valued at $180 each; 2 lighters at $150 each, and 10 skiffs, etc., at $30 each.

The statistics for 1897 were about the same as for 1896, but besides the seines, 10 gill nets were tried, 180 fathoms long, 4 fathoms deep, 5½-inch and 6-inch mesh, and valued at $150 each; and in addition to the *Marie G. Haaven*, another steamer, the *Herald*, was purchased. This was of 19.4 net tons burden, had a crew of 5, and was valued at $7,000.

The gill nets were not very successful, the failure being attributed to a too large mesh. Mr. Duncan thinks that 5 and 5½ inch mesh might have done better work. Gill nets have been tried all over Alaska and are now used with success only where the water is discolored; if the water is clear the fish see the net and the catch is small. In some places, where the waters are discolored, gill nets only are used. Of these places, there may be mentioned the mouths and approaches of the Unuk, Stikine, Taku, and Chilkat rivers in southeast Alaska, Copper River, Cook Inlet, and the Bering Sea district. Fish can be taken in gill nets in other places, but where the water is clear other fishing methods are far more successful.

The Metlakahtla cannery pays for redfish, according to locality, from $5.50 to $7 per 100; for cohoes, from $10 to $14 per 100, and for humpbacks, $1.50 per 100. These prices are unusually high. The redfish is, of course, the species most sought for, but as the canneries increase in number and the output becomes larger, more humpbacks and cohoes are packed to fill up the cannery quota.

—Moser, *Salmon Fisheries of Alaska*, pp. 66-67.

Opposite—*Labeling cans of salmon prior to shipment.*

JOHN N. COBB COLLECTION, UNIVERSITY OF WASHINGTON LIBRARY

Left—*Making boxes at Naknek Packing Company's cannery, July 1918. Wooden boxes for the entire pack were made by hand.* Below—*Salmon cooling after being cooked.* Bottom—*Naknek Packing Company facilities in early 1900's.*

COURTESY OF JOHN GILBERT

COURTESY OF HAROLD BRINDLE

From: *Pacific Fisherman Year Book*, 1918, p. 88.

Specifications for Shipping Cases*

By DR. A. W. BITTING.

*From National Canners' Association Bulletin No. 40.

Box-testing machine designed for the tests described. Obstructions of various kinds on the inside of the drum cause the boxes to slide, roll, and drop in different positions.

The following specifications for better containers have been approved by the joint committee representing the National Wholesale Grocers' Association, the National Association of Box Manufacturers, and the National Canners Association, and adopted by the latter association. These specifications represent the culmination of three years' work in the collecting of general information, and nearly two years' work in making tests of boxes under various conditions. The specifications recommended are based upon extensive tests, and upon more nearly accurate knowledge of the strength of the materials used, mode of construction, and hazards to which the packages will be subjected, than any previous box specifications.

The National Association of Box Manufacturers, cooperating with the Forest Products' Laboratory of the United States Department of Agriculture, have procured material of all kinds from which boxes are usually made, have subjected it to different treatment in storage and drying, have built the boxes, using the same variables in construction, and have submitted all to the same hazards. From the results obtained, the woods have been classified and a mode of construction given which will insure a good box; one which will carry the goods to their destination and protect them against any usual treatment which they may receive in transportation.

Photograph showing destruction of wood about the nail head, due to over-driving. The tests showed clearly that the strength of the box depends more upon the nailing than on the thickness of the material used.

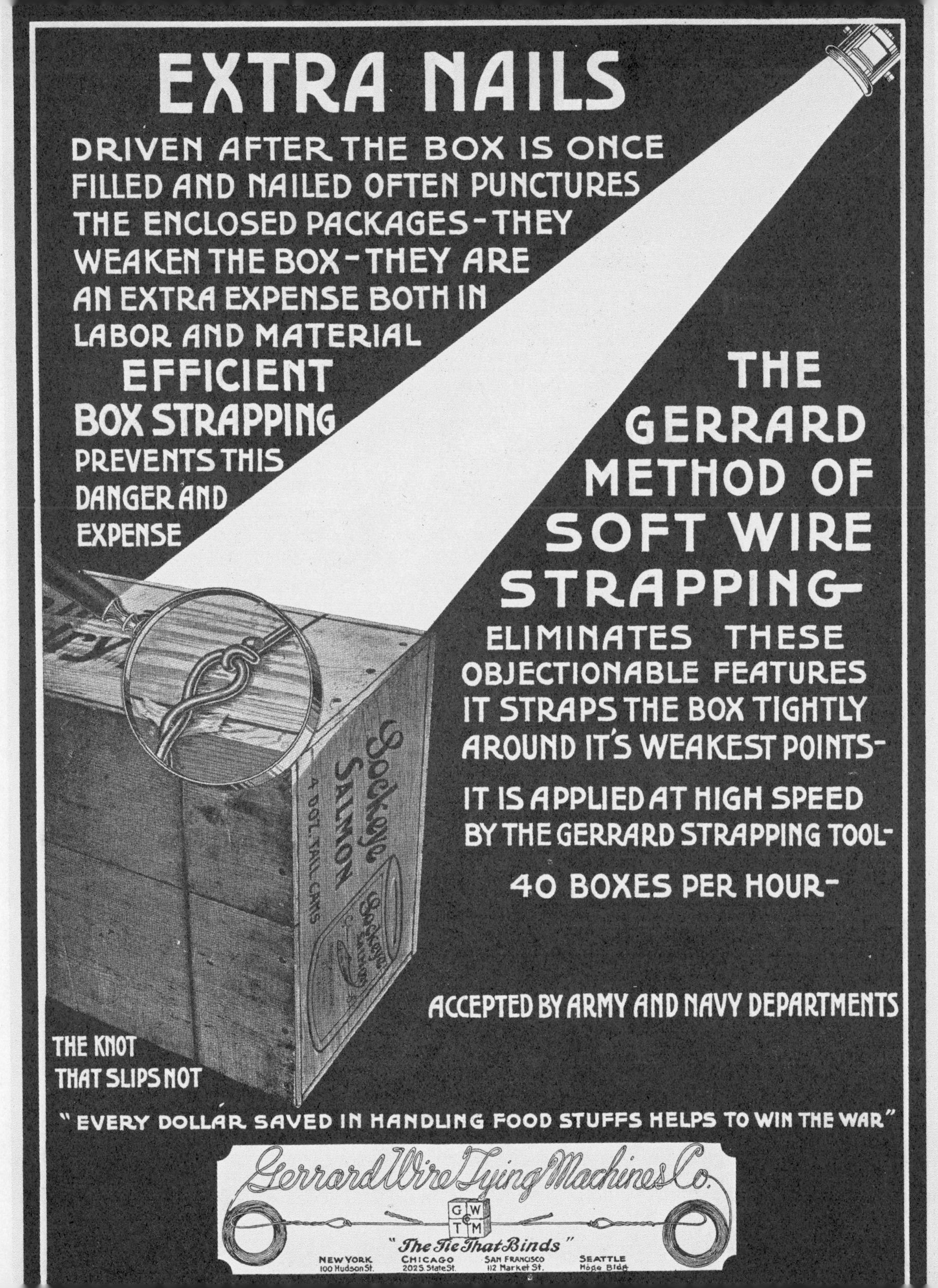

COURTESY OF WALTER YONKER

Canned salmon stacked and waiting casing prior to shipping.

REPRINTED FROM PACIFIC FISHERMAN,
NOVEMBER 1917, P. 20

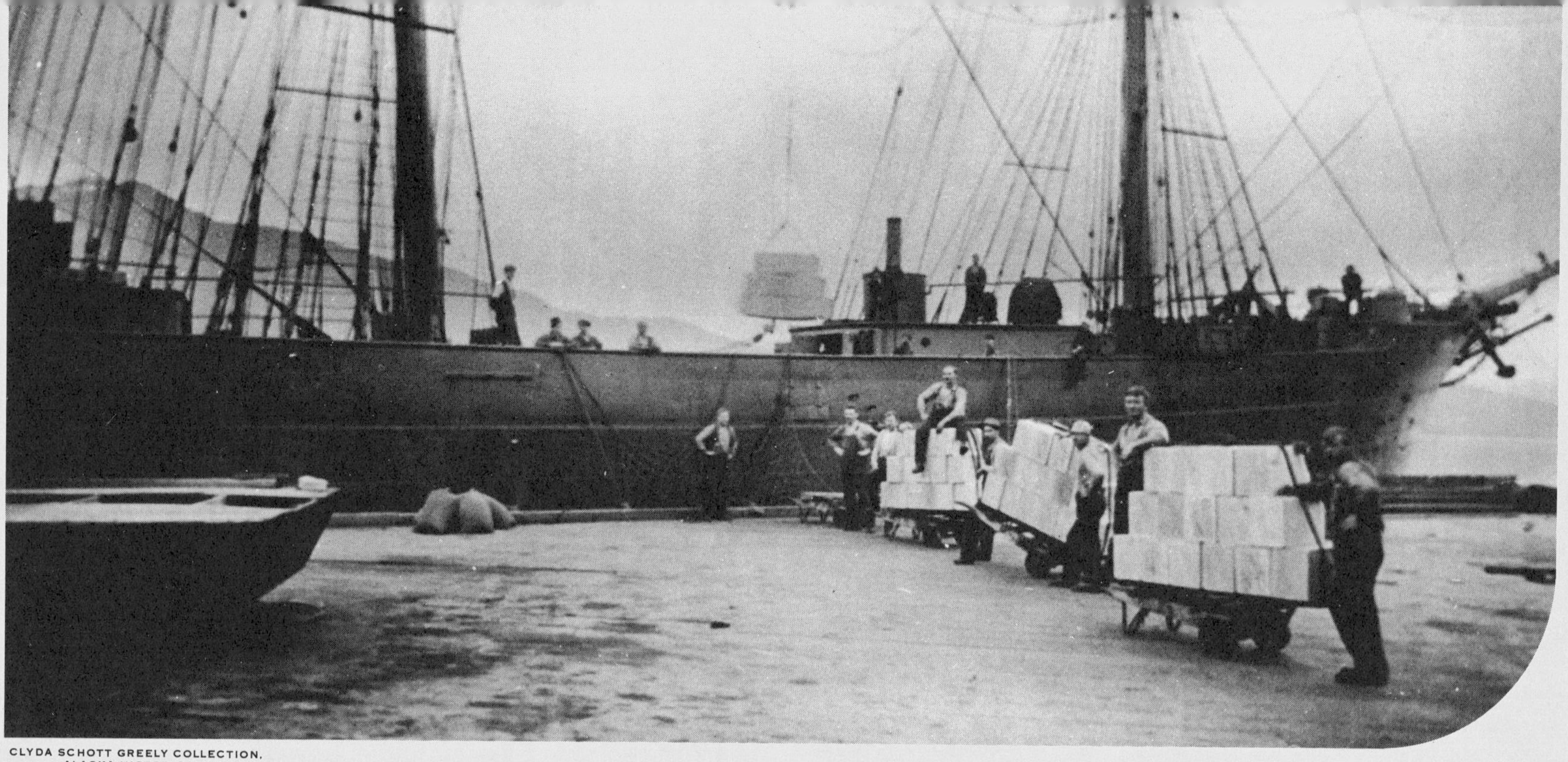

CLYDA SCHOTT GREELY COLLECTION,
ALASKA HISTORICAL LIBRARY

Above—*Loading salmon aboard the* Star of Greenland *at Wrangell Alaska Packers' Association cannery, 1914.* Right—*Handmade wooden boxes.*

COURTESY OF WALTER YONKER

REPRINTED FROM PACIFIC FISHERMAN, JANUARY 1917, P. 38

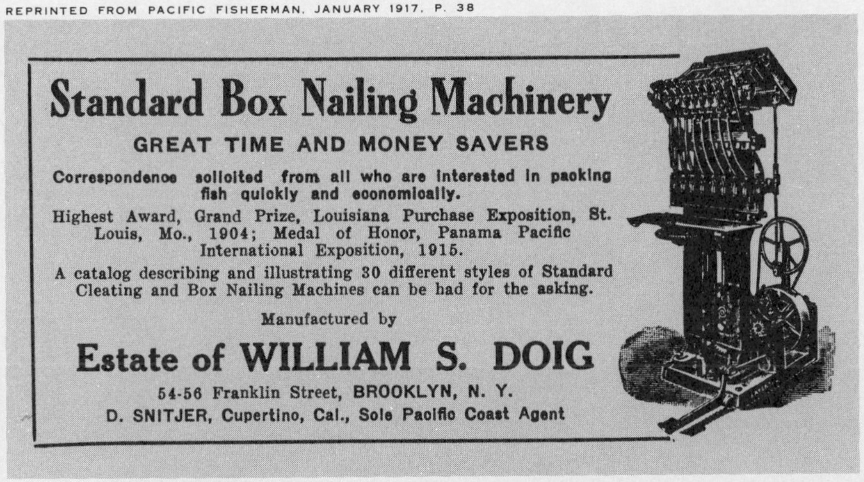

Right—*Fiber cases at Nakat Packing Corporation, Ketchikan, 1927. The wooden cases had become a thing of the past. Boxes had been made in Wrangell.*

COURTESY OF NATIONAL FISHERMAN

BOLLA COLLECTION, SAN FRANCISCO MARITIME MUSEUM

COURTESY OF HAROLD BRINDLE

Far left—*Moving cases of salmon on a lighter.* Near left—*Loading cases at Naknek Packing Company, Naknek, Alaska, 1926.*

C. W. BROWN COLLECTION, SAN FRANCISCO MARITIME MUSEUM

SS Nushagak *with scow ready to load cases of salmon for lightering to a nearby ship. Photo was taken in early 1900's.*

From: Moser, *Salmon Fisheries of Alaska*, pp. 173-177.

BERING SEA DISTRICT

The following brief histories and statistics of the Bering Sea canneries have been largely furnished by the Alaska Packers' Association, time not permitting a visit.

From the table of percentages of packs for the different districts, it will be seen that 35.2 per cent of the Alaska pack was made in Bering Sea in 1887. By the increase of canneries in other districts in 1888, and the still larger increase in 1889, the percentage of pack fell off and reached its lowest point (13.4 per cent) in 1892. From that date the percentage of pack has gradually increased until, in 1897, it reached 28 per cent of the whole Alaska pack, of which over nine-tenths was packed by the Alaska Packers' Association.

All the canneries in Bering Sea are located on the Nushagak, Kvichak, Naknek, and Ugashik rivers, all emptying into Bristol Bay.

NUSHAGAK RIVER

On the Nushagak there are four canneries, as follows: Arctic Packing Company, Nushagak Packing Company, Bristol Bay Canning Company, and Alaska Packing Company. There is also a saltery operated by C. E. Whitney & Co., which in 1897 salted 2,436 barrels, principally redfish.

continued

PHOTOS COURTESY OF JOHN GILBERT

Above—*The Native method of setting gill nets, on the beach near the Pacific Steam Whaling Company's cannery, Nushagak Bay.*
Left—*Arctic Packing Company, Naknek River, built in 1890. In 1893 it was purchased by the Packers' Association.*

Arctic Packing Company (APA) cannery at Nushagak Bay, 1890's.

COURTESY OF JOHN GILBERT

In 1883 the schooner *Neptune*, with a party, prospected for salmon on the Nushagak, and salted a large number. The same year cannery buildings were erected for the Arctic Packing Company, which was formed and commenced operations in 1884, making a pack of 400 cases that year. This was the first cannery operated in Bering Sea. It has made a pack every year to date, except in 1892, when it joined the pool of the Alaska Packing Association and was closed. In 1893 it became a member of the Alaska Packers' Association. The cannery is located at Kanulik, . . . about 3 miles above Fort Alexander. It has a capacity of 2,000 cases a day.

The Alaska Packing Company built a cannery on the western shore of the Nushagak, near the mouth, in 1886, and made a pack that year, as it has done every year since. It entered the pool of the Alaska Packing Association in 1892, and the Alaska Packers' Association in 1893. It is located near the village of Kanakanak, and has a capacity of 2,000 cases a day.

The Bristol Bay Canning Company built a cannery immediately above that of the Alaska Packing Company in 1886, and operated that year and every year to date. It entered the pool of the Alaska Packing Association in

continued

1892, and of the Alaska Packers' Association in 1893. It has a capacity of 2,000 cases per day.

The Nushagak Packing Company built a cannery on the eastern shore of the estuary at Nushagak in 1888, at a place called Stugarok, 11 miles below Fort Alexander. It was operated in 1888, 1889, 1890, and 1891; it was then closed and has not been operated since. It joined the pool of the Alaska Packing Association in 1892 and became a member of the Alaska Packers' Association in 1893. It is in such condition that it can be operated at short notice, and is held as a reserve. Its capacity is 1,500 cases per day.

These canneries are all owned and operated by the Alaska Packers' Association, and are under one local management. In 1897 they employed 144 white fishermen, 38 white cannery-hands, 306 Chinese, and 40 natives. Sixty gill nets were used, each 125 fathoms long by 24 meshes deep; 120 gill nets, each 75 fathoms long by 24 meshes deep; 4 traps with 300 feet leaders.

The following vessels were employed:

Rig.	Name.	Net tons.	Crew.	Value.
Steamer	Polar Bear	29	5	$12,000
Launch	Tyone	5	2	4,500
Do	Amy S	5	2	3,000
Do	Corinne	5	2	1,500
Ship	Sterling	1,637	Fishermen.	30,000
Do	Eclipse	1,535	(?)	Chartered.
Barkentine	Willie R. Hume	632	(?)	Do.
98 boats and lighters.				

The following gives the output of the three canneries for 1897:

Species.	Number of fish.	Number of cases.	Number of fish per case.	Barrels salted.
King salmon	18,139	5,823	3	32
Redfish	1,240,080	88,791	14	178
Cohoes	150,000	10,119	14.8	
Humpbacks	35,348	3,123	11.3	
Total		107,856		

continued

Source: Marsh, Millard C., and Cobb, John N. "The Fisheries of Alaska in 1908," *Report of the Commissioner of Fisheries for the Fiscal Year 1908 and Special Papers.* Washington: Government Printing Office, 1910, pp. 13-14.

PLANTS IN OPERATION.

Following is a list of the plants operated in Alaska during the season of 1908:

Name.	Location.
Southeast Alaska:	
Canneries—	
Alaska Packers Association	Loring, Wrangell, and Pyramid Harbor.
Northwestern Fisheries Co	Quadra, Hunter Bay, Santa Ana, and Dundas Bay.
Astoria and Puget Sound Packing Co	Excursion Inlet.
William Duncan	Metlakahtla.
Fidalgo Island Packing Co	Ketchikan.
C. A. Burckhardt & Co	Yes Bay.
Gorman & Co. (Incorporated)	Kasaan Bay.
F. C. Barnes	Lake Bay.
Shakan Salmon Co	Shakan.
North Pacific Trading and Packing Co	Klawak.
Pillar Bay Packing Co	Pillar Bay.
Pacific Coast and Norway Packing Co	Petersburg.
John L. Carlson	Taku Harbor.
George T. Myers & Co	Sitkoh Bay.
Thlinket Packing Co	Funter Bay.
Pacific-American Fisheries	Excursion Inlet.
Columbia Canning Co	Haines.
Yakutat and Southern Railway Co	Yakutat.
Salteries, etc.—	
Mrs. A. E. King	Sunny Point.
Fred Brockman	Sarkaar.
Rasmus Engee	Petersburg.
James Thompson	Skowl Arm.
K. J. Johansen	Ideal Cove.
H. E. Heckman	Loring.
Louis Peterson	Threemile, Kupreanof Island.
Knutsen Brothers	West Point.
John Baronovitch	Skowl Arm.
Peter Jorgenson	Wrangell Narrows.
W. C. Waters	Holbrook.
John H. Mantle	Olive Bay.
Forss & Skodlung	Thorne Bay.
Craig Millar	Hunter Bay.
Peter Sommers	Petersburg.
Alex. S. Millar	Nakat Inlet.
Beauclerc Salting Co	Beauclerc Bay.
S. E. Robertson	Moira Sound.
Globe Fishing and Packing Co	Dall Island.
Knute Hauge	Wrangell Narrows.
M. E. Lane	Prince of Wales Island.
H. Appleton	Gasoline sloop Capella.
W. A. Thompson, John Wilson, C. J. Parker	Gasoline vessels Annie M and Eurus

continued

COURTESY OF JOHN GILBERT

Canning plant at Point Roberts Packing Company (APA) at Koggiung, Kvichak Bay. Photo was taken looking southwest in early 1900's.

PLANTS IN OPERATION. (continued)

Name.	Location.
Engelbr. Wiese	Stikine River and Ketchikan.
J. Lindenberger (Incorporated)	Ketchikan, Klawak, and Douglass.
J. T. Field	Petersburg.
A. H. Sonsthagen	Cape Fanshaw.
Malcolm Campbell	Alsek River.
Schooner "C. Seward"	Juneau.
Schooner "Alcedo"	Do.
Central Alaska:	
Canneries—	
Alaska Packers Association	Chignik, Karluk, Alitak, and Kasilof.
Northwestern Fisheries Co	Orca, Uyak, and Chignik.
Salteries, etc.—	
Alaska Commercial Co	Kodiak.
Blodgett & Blinn	Do.
J. A. Herbert	English Bay.
San Juan Fishing and Packing Co	Kenai.
Thin Point Packing Co	Thin Point and Simeonof Island.
B. Omundsen	Osonio.
Western Alaska:	
Canneries—	
Alaska Packers Association	Naknek River, Ugaguk River, Nushagak Bay, and Kvichak Bay.
Northwestern Fisheries Co	Nushagak Bay.
Red Salmon Canning Co	Ugashik River.
North Alaska Salmon Co	Nushagak Bay, Lockonok, Kvichak River, and Ugaguk River.
Naknek Packing Co	Naknek River.
L. A. Pedersen	Kvichak Bay.
Alaska-Portland Packers Association	Nushagak Bay.
Columbia River Packers Association	Do.
Alaska Fishermen's Packing Co	Do.
Alaska Salmon Co	Wood River.
Salteries, etc.—	
Alaska Packers Association	Ugashik River.
Lagoon Salmon Co	Nelsons Lagoon.
Peter M. Nelson	Igushik River.
Nelson, Olsen & Co	Kvichak Bay.

Source: Marsh and Cobb, *Fisheries of Alaska, 1908,* pp. 13-14.

Continued

continued from p. 37.

Of the above, the Bristol Bay Canning Company packed 34,117 cases, the Alaska Packing Company 37,849 cases, and the Arctic Packing Company 35,890 cases.

The traps are used in Wood River, which empties into the estuary above the canneries on the western bank; the gill nets are used in the estuary where the water is muddy. No seines are used.

KVICHAK RIVER

To the eastward of the Nushagak and emptying into the head of Bristol Bay is the Kvichak, with one cannery and one saltery.

The Prosper Fishing Company established a saltery at the mouth of the Kvichak in 1894 and operated it that year and the following. It was sold in 1896 to the Alaska Packers' Association and closed.

Under the name of Point Roberts Packing Company the Alaska Packers' Association established and operated a saltery at Koggiung on the Kvichak River in 1894, and built a cannery the following year, utilizing in its construction the available machinery from the cannery of the Central Alaska Company at Thin Point. It was first operated in 1896, and also packed in 1897. It is said to have a capacity of 2,000 cases per day. In 1897 the company employed 65 white fishermen, 10 white cannery-hands, 150 Chinese, and 25 natives. Twenty-five gill nets were used, each 60 fathoms long by 30 meshes deep, and 4 traps, with the inside leaders 300 feet long and outside leaders 250 feet long.

The following vessels were employed:

Rig.	Name.	Net tons.	Crew.	Value.
Steamer	President	238	8	$49,000
Launch	Northern Light	5	2	3,000
Do	J. W. Clark	5	2	2,000
Ship	Bohemia	1,529	Fishermen.	30,000
Bark	W. W. Case	556	Fishermen.	10,000
32 boats, lighters, and scows.				

The following is the pack for 1897:

Species.	Number of fish.	Number of cases packed.	Number of fish per case.	Barrels salted.
King salmon	345	126	2.7	220
Redfish	760,652	55,382	13.7	1,489

continued

Schooner Prosper *and steamer* Sayak *at Point Roberts Packing Company (APA) canning plant at Koggiung in early 1900's.*

COURTESY OF JOHN GILBERT

COURTESY OF JOHN GILBERT

A Native barabara *(sod house) and storehouse, near the Naknek River.*

NAKNEK RIVER

Naknek River empties into the upper part of Bristol Bay, about 20 miles below Koggiung. There are two canneries at its mouth.

The Arctic Packing Company built and operated a saltery at Naknek in 1890 and sold it to the Alaska Packers' Association in 1893. The association in 1894 built a cannery at the same place, utilizing the machinery of the Thin Point Packing Company in its construction. It was operated in 1895, 1896, and 1897, and has a capacity of 1,800 cases per day. Salmon are also salted here.

In 1897 the company employed 45 white fishermen, 10 white cannery-hands, 102 Chinese, and 10 natives. They used 11 gill nets, each 70 fathoms long by 24 meshes deep; 34 gill nets, each 70 fathoms long by 22 meshes deep; 3 traps with leads of 175 feet. The vessels employed were the launch *Ralph L.*, of 5 tons, with a crew of 2, and valued at $4,500; the bark *Merom*, of 1,159 net tons, with a crew of fishermen, valued at $16,000; the schooner *Prosper*, of 230 net tons, with a crew of fishermen, valued at $15,000; also 30 boats and lighters. The bark *Merom* made one trip on account of Karluk.

The following gives the pack for 1897:

Species.	Number of fish.	Number of cases packed.	Number of fish per case.
King salmon	902	180	5
Redfish	429,655	34,496	12.4

In 1890 Mr. L. A. Pederson established and operated a small saltery on the northern shore of the Naknek River near its mouth, and in 1894 the Naknek Packing Company was incorporated under the laws of the State of California, a cannery being built on the site of the saltery. The first pack was made in 1895, and operations were continued in 1896 and 1897. Salting is also carried on by

continued

COURTESY OF JOHN GILBERT

Left—*Arctic Packing Company facility at Naknek, 1896, is now (NN) canning company of Alaska Packers' Association.* Left below—*Freehand sketch of mouth of Naknek River, Bristol Bay, from* Alaska Salmon Investigation in 1900. Below—*Retubing a boiler at a Naknek cannery. Workers pictured are unknown, but man at left was chief mechanic. Skilled mechanics and journeymen, many from Europe, kept the canneries operating. Emergency parts were made by hand. The "blacksmith" was an important man.*

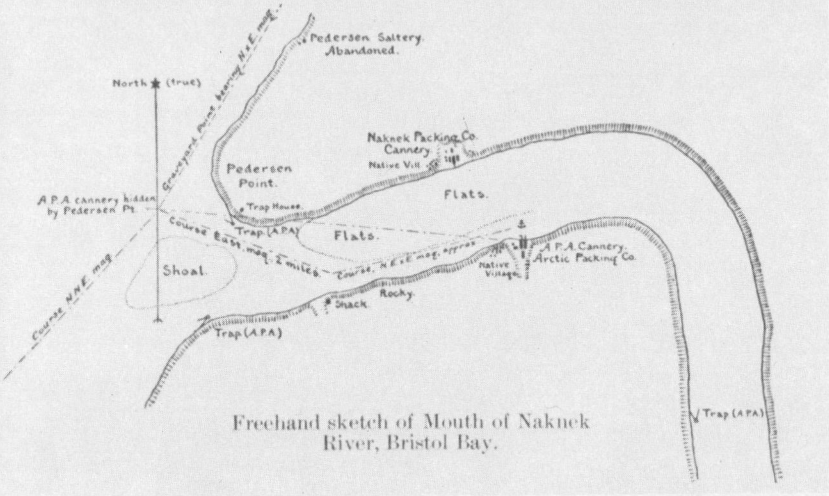

COURTESY OF JOHN GILBERT

C. W. BROWN COLLECTION, SAN FRANCISCO MARITIME MUSEUM

PHOTOS COURTESY OF JOHN GILBERT

Top—*Ukala (dried salmon) racks near the Naknek River cannery of Arctic Packing Company (APA). On the left is a tent typical of those used by the Bering Sea Eskimos and Aleuts. Photo was taken about 1900.* Bottom—*A bidarka and a kayak on the beach at Ugashik River, circa 1900. Skin boats such as these were used by the Natives to travel from villages to the canneries.*

this company. In 1897 an additional saltery was built on the shore of Bristol Bay about 2 miles above the mouth of the Naknek. The cannery has a capacity of 1,500 cases per day.

In 1897 the company employed 40 white fishermen and salters, 10 white cannerymen, 75 Chinese, and about 10 natives. The following was the equipment: 27 gill nets, each 100 fathoms long by 20 meshes deep, 6¼-inch mesh, valued at 40 cents per fathom; 3 gill nets, each 100 fathoms long by 20 meshes deep, 8½-inch mesh, valued at 40 cents per fathom; also 2 seines each 60 fathoms long, 100 meshes deep at bunt and 60 at ends, 3-inch mesh, worth $1.25 per fathom (used to empty trap); 1 double trap leading out 100 fathoms from beach in front of cannery, with two wings from the pot, 50 fathoms each in length, one up and the other down stream; width of pot, 20 feet; value, $600.

The vessels and boats were the launch *Emilia*, of 5 tons, with a crew of 2, and valued at $1,200; the bark *B. P. Cheney*, of 1,200 tons, with a crew of fishermen, and valued at $14,000; 1 lighter at $700; 6 lighters at $300 each; 1 pile-driver with engine valued at $300; 20 double-ended gill-net boats at $90 each.

In 1896 the company employed 36 white fishermen and salters, 6 white cannery-hands, 62 Chinese, and 10 natives. The remaining statistics are about the same as for 1897, except that the large lighter valued at $700 was new in 1897, and only 16 gill-net boats and 22 gill nets were used. The bark *Ferris S. Thompson*, of 514 tons, was chartered for transport.

The pack for 1897 consisted of 216,500 redfish, of which 215,000 were taken in gill nets and 1,500 in traps, making 18,000 cases, or 12 fish to the case. 16,000 of these fish were taken outside, the remainder in the river. Packing began June 30, and finished August 1. At the saltery 501 barrels of redfish were salted, 48 fish to the barrel,

continued

between July 1 and 14, and 99 barrels at the cannery.

In 1896 8,600 cases were packed from July 2 to August 2, all redfish, averaging 12 to the case. 175 barrels of redfish were salted to order.

The following redfish, running 48 to the barrel of 200 pounds, and all taken in the lower Naknek River, have been salted by this company since the station was opened: 225 barrels in 1890; 450 barrels in 1891; 1,100 barrels in 1892; 2,600 barrels in 1893; 2,630 barrels in 1894; 200 barrels in 1895.

EGEGAK FISHING STATION

Egegak fishing station, also called Igagik and Ugaguk, is about 34 miles south of Naknek. In 1895 the Alaska Packers' Association established and operated a saltery at the mouth of the river, and have salted there each year since. In 1897 the company employed 10 white fishermen, 16 white saltery-hands, and 6 natives. Five gill nets were used, each 75 fathoms long by 26 meshes deep, and 2 traps with leads of 800 feet. The vessels and boats were the launch *Minnie*, of 5 tons, with a crew of 2, and valued at $1,600, and 9 boats and lighters. The three-mast schooner *Premier*, of 292 tons, with a crew of fishermen, and valued at $15,000, was used as a transport.

In 1897 this company salted 257 king salmon, making 15 barrels, and 203,458 redfish, making 3,574 barrels.

UGASHIK RIVER

The Ugashik, or Sulima River, as it is sometimes called, is 70 miles south of Naknek, and is the most southerly station on Bristol Bay. There are two salteries and two canneries on this river.

The Bering Sea Packing Company, a branch of the Alaska Improvement Company, but a separate corporation, built a cannery at Ugashik in 1891, and operated it that year. It was closed in 1892 and in 1893, and operated in 1894, 1895, and 1896. It was neither in the pool of the Alaska Packing Association of 1892, nor did it join the

continued

OPENING PRICES SINCE 1897.

1897.

	Talls
Columbia River Chinook	$1.05
Puget Sound Sockeye	.80
Alaska Red	.90
Alaska Pink	.65

1898.

	Talls
Columbia River Chinook	$1.05
Puget Sound Sockeye	.80
Alaska Red	1.00
Alaska Pink	.65

1899.

	Talls
Columbia River Chinook	$1.25
Puget Sound Sockeye	1.10
Alaska Red	1.00
Alaska Pink	.67½

1900.

	Talls
Columbia River Chinook	$1.60
Puget Sound Sockeye	1.10
Alaska Red	1.10
Alaska Pink	.75

1901.

	Talls
Columbia River Chinook	$1.50
Puget Sound Sockeye	.95
Alaska Red	1.25
Alaska Pink	.75

1902.

	Talls
Columbia River Chinook	$1.35
Puget Sound Sockeye	1.00
Alaska Red	.95
Alaska Pink	.65

1903.

	Talls	Flats	Halves
Puget Sound Sockeye	$1.50	$1.60	$ .90
Columbia River Chinook	1.35	1.45	.85
Alaska Red	1.30	.....	.....
Alaska Pink	.50	.....	.....

1904.

	Talls	Flats	Halves
Columbia River Chinook	$1.45	$1.15	$ .90
Puget Sound Sockeye	1.55	1.65	.95
Alaska Red	1.30	.....	.....
Alaska Pink	.70	.....	.....

1905.

	Talls	Flats	Halves
Columbia River Chinook	$1.45	$1.55	$ .90
Puget Sound Sockeye	1.35	1.50	4.00
Alaska Red	1.00	.....	.....
Alaska Pink	.70	.....	.....

1906.

	Talls	Flats	Halves
Columbia River Chinook	$1.50	$1.60	$1.00
Puget Sound Sockeye	1.45	1.60	1.00
Alaska Red	.95	.....	.....
Alaska Pink	.75	.....	.....

1907.

	Talls	Flats	Halves
Columbia River Chinook	$1.65	$1.75	$1.05
Puget Sound Sockeye	1.65	1.75	1.10
Alaska Red	1.15	.....	.....
Alaska Pink	.80	.....	.....

1908.

	Talls	Flats	Halves
Columbia River Chinook	$1.65	$1.75	$1.05
Puget Sound Sockeye	1.60	1.75	1.05
Puget Sound Pink	.75	.80	.....
Puget Sound Cohoe	1.05	1.15	.75
Alaska Red	1.15	.....	.....
Alaska King	1.05	.....	.....
Alaska Cohoe	1.00	.....	.....
Alaska Pink	.70	.....	.....
Alaska Chum	.70	.....	.....

1909.

	Talls	Flats	Halves
Columbia River Chinook, Fancy	$1.65	$1.75	$1.05
Puget Sound Sockeye	1.35	1.50	1.00
Alaska Red	1.15	1.35	.85
Alaska King	1.10	.....	.....
Alaska Cohoe	1.05	1.20	.70
Alaska Pink	.60	.....	.....
Alaska Chum	.57½	.....	.....

1910.

	Talls	Flats	Halves
Columbia River Chinook, Fancy	$1.75	$1.90	$1.10
Puget Sound Sockeye	1.65	1.80	1.10
Alaska Red	1.35	1.50	1.00
Alaska King	1.35	.....	.....
Alaska Pink	.80	.....	.....
Alaska Chum	.77½	.....	.....
Medium Red and Cohoe	1.25	1.40	.80

1911.

	Talls	Flats	Halves
Columbia River Chinook, Fancy	$1.95	$2.00	$1.30
Puget Sound Sockeye	1.95	2.00	1.30
Alaska Red	1.60	1.75	1.12½
Alaska Medium Red	1.45	1.65	1.00
Alaska King	1.80	2.00	1.12½
Pink	1.00	1.15	.80
Chum	.95	1.05	.75

1912.

	Talls	Flats	Halves
Chinook	$1.95	$2.00	$1.25
Sockeye	1.95	2.00	1.30
Alaska Red	1.40	1.60	1.15
Alaska Medium Red	1.15	1.25	.80
Alaska King	1.40	1.60	1.15
Pink	.65	.65	.55
Chum	.62½	....	.50

1913.

	Talls	Flats	Halves
Chinook	$1.95	$2.00	$1.25
Sockeye	1.50	1.65	1.05
Alaska Red	1.15	1.35	.95
Alaska Medium Red	.85	1.00	.70
Alaska King	1.00	1.15	.90
Pink	.65	.80	.55
Chum	.55	.70	.50

1914.

	Talls	Flats	Halves
Chinook	$1.95	$2.10	$1.25
Sockeye	1.95	2.15	1.35
Alaska Red	1.45	1.80	1.10
Medium Red	1.15	1.35	.82½
Alaska King	1.40	....	1.10
Pink	.90	1.00	.70
Keta, or Chum	.85	.95	.65

Source: *Pacific Fisherman Year Book*, 1914, p. 57.

Alaska Packers' Association in 1893. It was purchased by the latter organization in the spring of 1897 and closed. It has a capacity of 1,200 cases per day, and at present is held as a reserve.

A saltery was established and operated at Ugashik in 1893 by the Alaska Packers' Association, and continued as such in 1894 and 1895. In the latter year the association built a cannery, utilizing in its construction the available machinery from the cannery of the Russian-American Packing Company at Afognak. It was operated in 1896 and 1897, and has a capacity of 1,800 cases per day.

At the Ugashik fishing station in 1897, 59 white fishermen, 9 white cannery-hands, 102 Chinese, and 24 natives were employed. Twenty-one gill nets were used, each 75 fathoms long by 24 meshes deep; also, one trap 150 feet inside leader, 80 feet outside leader, and 1 trap 250 feet inside leader, 80 feet outside leader.

The vessels and boats were the steamer *Thistle*, of 55 tons, with a crew of 6, and valued at $25,000; the launch *Cathie K*, of 5 tons, with a crew of 2, and valued at $2,500; the bark *Coryphene*, of 771 tons, chartered; also 53 boats and lighters.

The pack in 1897 consisted of 259 king salmon, making 11 cases (nearly all consumed fresh); 463,698 redfish, making 38,261 cases, or 12 to the case. 138 barrels were salted.

Mr. C. A. Johnson established a saltery at Ugashik in 1889, and has salted every year to date. The names of Norton, Teller & Co., and Metson & Co. frequently appear in connection with this saltery in the lists. They are simply the saltery agents.

Mr. Charles Nelson established a saltery at Ugashik in 1893, and operated it that year and in 1894. In 1895 it was sold to the Alaska Packers' Association and closed.

TOGIAK AND PORT MOLLER

Togiak is on Togiak Bay, to the westward of Bristol Bay, between Cape Constantine and Cape Newenham. In 1895 the Alaska Packers' Association established a saltery here, which was operated in 1895 and 1896, and then abandoned.

At Port Moller, on the northern side of the peninsula, in longitude 160°40' W., Captain Herendeen salted salmon in 1886 or 1887.

WAGES IN BERING SEA DISTRICT

The Chinese and fishermen's contracts for Bering Sea are similar to those referred to in other localities. The following for one cannery will give a fair idea of their value:

The Chinese received free transportation and furnished quarters (without bedding), fuel, water, and salt. They are paid 42 cents per case. The boss receives $50 per month in addition to his lay, and the tester receives the same.

The fishermen receive $25 for working the vessel to the cannery and a like amount for taking her back. Two men form a gill-net crew, and the boat is paid 2 cents per fish. They are furnished everything except clothing, from the time they go on board the vessel until discharged.

Saltery gangs are paid at the rate of $30 per month and 15 cents a barrel (of 200 pounds) for the gang collectively.

The beach gang receives $40 per month per man, and, if detailed for fishing, in addition to this pay, receives ½ cent per fish. A cook and boy are provided to run the mess.

The natives are paid in various ways, but average from $1, the lowest, to $1.50 per day.

—Moser, *Salmon Fisheries of Alaska,* pp. 173-177.

From: *Pacific Fisherman*, January 1914, Vol. XII, No. 1, p. 55.

Investment in the Salmon Canning Industry in 1913.

ITEMS	SOUTHEAST ALASKA		CENTRAL ALASKA		WESTERN ALASKA		TOTAL	
	No.	Value	No.	Value	No.	Value	No.	Value
Canneries	42	$3,946,518	14	$1,416,880	23	$3,136,351	79	$8,499,749
Working capital		4,298,930		2,147,600		5,374,900		11,821,430
Wages paid		1,946,736		893,250		2,516,956		5,356,942
Vessels:								
Power vessels over 5 tons	111	611,149	44	471,740	55	924,120	210	2,007,009
Net tonnage	2,118		1,790		4,822		8,730	
Launches under 5 tons	39	47,860	16	20,527	18	27,851	73	96,238
Sailing	8	214,000	11	412,236	34	864,932	53	1,491,168
Net tonnage	10,131		19,844		41,856		77,831	
Boats, sail and row	636	45,241	442	41,832	1,056	207,688	2,134	294,761
Lighters	242	121,112	188	119,666	145	163,852	575	404,630
Pile drivers	38	124,363	31	86,316	19	41,375	88	252,054
Apparatus:								
Haul seines	60	14,200	47	17,385			107	31,585
Fathoms	7,419		11,449				18,868	
Purse seines	184	81,767	6	4,221	6	4,164	196	90,152
Fathoms	33,432		1,130		1,440		36,002	
Gill nets	257	30,477	183	17,064	1,624	192,883	2,064	240,424
Fathoms	62,175		19,495		234,981		316,651	
Traps, driven	127	372,039	84	271,538	13	19,369	224	662,946
Traps, floating	36	92,553					36	92,553
Dip nets	18	29					18	29
Total		11,946,974		5,920,255		13,474,441		31,341,670

COURTESY OF JOHN GILBERT

Labor employed, 1900.

Name of company and location of cannery.	Fishermen.		Other employees.			Total employees.
	White.	Native.	White.	Native.	Chinese.	
Arctic Packing Co. / Alaska Packing Co. / Bristol Bay Canning Co. (A. P. A.), Nushagak Bay	215		66	75	450	806
Pacific Steam Whaling Co., Nushagak Bay	56		10	35	100	201
Alaska Fishermen's Packing Co., Nushagak Bay	55		7	35	93	190
Point Roberts Packing Co. / Kvichak Packing Co. (A. P. A.), Kvichak Bay	154		25 / 13	25 / 8	[1]300 / [2]147	672
Arctic Packing Co. (A. P. A.), Naknek River	58		54	20	140	272
Naknek Packing Co., Naknek River	60		12	11	131	214
Egegak Packing Co. (A. P. A.), Egegak River	39		16	10	80	145
Ugashik Fishing Station (A. P. A.), Ugashik River	64		24	20	140	248
Bristol Packing Co., Ugashik River	27		6	10	48	91
Chignik Bay Co. (A. P. A.), Chignik Lagoon	63		[3]20	2	90	175
Pacific Steam Whaling Co., Anchorage Bay, Chignik	60		10	4	60	134
Hume Bros. & Hume, Anchorage Bay, Chignik	52		10	6	58	126
Arctic Packing Co. (A. P. A.), Olga Bay, Alitak	40		9	10	59	118
Karluk Packing Co. (A. P. A.), Karluk / Alaska Improvement Co. (A. P. A.), Karluk / Uganuk Fishing Station (A. P. A.), Uganuk Bay	171	13	43	8	263	498
Pacific Steam Whaling Co., Uyak Bay	64	26	6	4	56	156
Hume Bros. & Hume, Uyak Bay	52	9	12		56	129
Alaska Salmon Association, Chuitna River	39		16		51	106
Pacific Steam Whaling Co., Kenai	40	10	6	6	60	122
Arctic Fishing Co. (A. P. A.), Kussilof River	45	10	10		100	165
Pacific Packing Co. (A. P. A.), Odiak	60	6	30	6	60	162
Pacific Steam Whaling Co., Orca	60		13	2	60	135
Western Fisheries Co., Dundas Bay	9	26	5	[4]26	30	96
Pyramid Harbor Packing Co. (A. P. A.), Pyramid Harbor	102	[5]10	12	[6]16	86	226
Chilkoot Packing Co., Chilkoot Inlet	24	[7]8	9	17	28	86
Taku Packing Co., Taku Inlet	30	14	7	2	19	72
Taku Fishing Co., Port Snettisham	16	20	4		25	65
Icy Strait Packing Co., Petersburg, Wrangell Narrows	26	28	7	[8]42	50	153
Royer-Warnock Packing Co., Beecher Pass	([9])	([9])	2		[10]11	13
Glacier Packing Co. (A. P. A.), Point Highfield	24	150	7	4	86	271
Thlinket Packing Co., Gerard Point	35	30	9	[11]7	49	130
Fidalgo Island Canning Co., Ketchikan	34	[12]41	7	[13]50	50	182
Alaska Salmon Packing and Fur Co. (A. P. A.), Loring	100	150	20	40	120	430
Boston Fishing and Trading Co., Yes Bay	15	15	5	15	61	111
Quadra Packing Co., Mink Arm, Boca de Quadra	24	40	5	[6]25	43	137
Metlakahtla Industrial Co., Metlakahtla		[14]24		174	174	372
North Pacific Trading and Packing Co., Klawak		55	16	[15]13	46	130
Pacific Steam Whaling Co., Hunter Bay	21	([16])	10		90	121
Total employees	1,934	685	543	728	3,570	7,460

[1] Includes a few Japanese.
[2] Chinese and Japanese.
[3] Includes 3 coal miners.
[4] Includes 6 Klootchmen.
[5] Purchased fish from 224 natives.
[6] Klootchmen.
[7] Purchased fish from 16 natives in addition.
[8] Includes 28 Klootchmen.
[9] Purchased all fish.
[10] 10 Japanese; 1 Chinaman.
[11] Includes 5 Klootchmen.
[12] Includes 3 Japanese.
[13] Includes 40 Klootchmen.
[14] Purchased fish from 38 natives in addition.
[15] 3 Bucks; 10 Klootchmen.
[16] Purchased fish from 121 natives.

F. C. B. 1901—21

INSPECTION OF CANNERIES

During the last year or so there has developed a feeling among various cannerymen that a Government inspector should be stationed at each cannery to pass upon the condition of fish packed, also to see that sanitary features are fully observed, in manner similar to the federal inspection of meat packing houses and products thereof. The theory of extending such inspection to Alaska is sound, but no funds are available for the employment of a force of 50 or more competent inspectors required to put the plan into effect. The managers of at least two canneries realizing the practical benefits of announcing the sale of goods packed under federal supervision, have stated their willingness to pay the salary and expenses of an authorized and impartial Government inspector stationed at each of their plants. There are practical difficulties in the way of accepting pecuniary aid from private enterprise, therefore provision by Congress of a personnel and funds is the proper way to effectuate the plan.

CANNERIES OPERATED

During 1913, there were 42 canneries in operation in Southeast Alaska, 9 less than in 1912; 14 in Central Alaska, the same as 1912, while in Western Alaska there were 23 canneries in operation, a gain of one over the previous year. Thus the totals are 87 for 1912 and 79 for 1913.

Persons Engaged in the Salmon Canning Industry in 1913.

OCCUPATIONS AND RACES	SOUTHEAST ALASKA	CENTRAL ALASKA	WESTERN ALASKA	TOTAL
Fishermen:				
Whites	584	752	2,027	3,363
Indians	1,157	220	45	1,422
Japanese	2	.....	.....	2
Miscellaneous	2	.....	.....	2
Total	1,745	972	2,072	4,789
Shoresmen:				
Whites	1,157	474	1,520	3,151
Indians	1,431	265	475	2,171
Chinese	865	328	857	2,050
Japanese	787	252	571	1,610
Miscellaneous	336	316	1,257	1,909
Total	4,576	1,635	4,680	10,891
Transporters:				
Whites	293	171	360	824
Indians	2	6	1	9
Total	295	177	361	833
Grand total.				
Whites	2,034	1,397	3,907	7,338
Indians	2,590	491	521	3,602
Chinese	865	328	857	2,050
Japanese	789	252	571	1,612
Miscellaneous	338	316	1,257	1,911
Total	6,616	2,784	7,113	16,513

Filipinos, Mexicans, Negroes, Porto Ricans, and so forth are included under the miscellaneous head.

From: *Pacific Fisherman*, September 1917, p. 25.

PATENT FOR NEW FISH TRAP.

Walter J. Larson, of Wrangell, Alaska, recently received a patent on a new type of floating fish trap invented by him. The most notable features of the invention are: A means of positively distending and holding distended the vertical webs of the trap, and preventing their displacement by currents or rough water; a form of trap permitting the use of two spillers, which are operated independently of each other and of the pot; an easily adjustable means of escape for the fish at times when it is not desired to operate the trap; a means of anchoring the floating trap which will hold it securely in position and will at the same time produce a minimum of strain upon the various parts of the trap; a device for holding the web of the leads distended and vertical without the use of heavy weights; and in general, the strong, compact character of the trap, which will not easily be broken up by rough weather.

From: *Pacific Fisherman*, February 1917, p. 16.

SAN JUAN FISHING & PACKING COMPANY BUILDS CANNERY AT SEWARD.

The San Juan Fishing & Packing company of Seattle has just closed a 20-year lease on a tract of land controlled by the government railroad at Seward, with a water frontage of 300 feet, and is starting work on a one-line salmon cannery and a small cold storage plant with suitable dockage. H. O. Roberts, formerly manager of the Pacific Fisheries, Ltd., of Prince Rupert, Canadian branch of the San Juan company, has taken the management of the company's operations at Seward where he arrived recently.

The terms of the lease require an expenditure of $100,000 on improvements this year, and the cannery will be ready for operations about the first of June. The cold storage plant will be used for halibut, bait, and miscellaneous fresh fish. Mr. Roberts states that the company has prospected the waters in the vicinity for halibut, etc., and will operate three purse seine boats and two halibut boats from that station during the coming season, and expects to handle the catches of independent fishermen.

PHOTOS FROM JOHN N. COBB COLLECTION, UNIVERSITY OF WASHINGTON LIBRARY

Above—*Seven canneries operated on the Karluk spit.* Below—*Gorman & Company and sawmill at Kasaan, Prince of Wales Island, 1908.*

From: Marsh and Cobb, *Fisheries of Alaska, 1908,* pp. 12-15.

THE SALMON INDUSTRY

As a whole the season of 1908 was the best ever known in the Alaska salmon industry. There were, however, marked fluctuations in the various regions as compared with previous years.

In central Alaska the season was fair, but not so good as in 1907, the catch of red salmon falling off from 6,637,860 fish in 1907 to 5,507,615 fish in 1908. The run in the Chignik was excellent, as usual, but the Karluk run was later in arriving and smaller in quantity than in 1907. The runs in Prince William Sound and Cook Inlet were fair in size.

In western Alaska, with the exception of the Ugashik River, the Bristol Bay region had a very large run of red salmon, especially in Nushagak Bay and the Kvichak and Naknek rivers. From these rivers fish were transferred to canneries on Nushagak Bay and to Ugaguk and Ugashik; 2,597 king salmon and 132,286 red salmon were lost in transit between the Kvichak River and Nushagak Bay and 86,070 red salmon were lost in transit from the Kvichak River to Koggiung. On the Naknek one cannery's boats were on limit for twenty days continuously, the limit remaining until the close of fishing. Fishing boats often procured their maximum in two hours' fishing.

The cannery of the Alaska Packers' Association at Coffee Creek, burned in 1906, was in process of rebuilding during 1908 and will be ready for operation in the season of 1909. Machinery will be installed from one of the company's canneries at Ugashik. Four traps were operated in Kvichak River and Bay, two by the Alaska Packers' Association and two by the North Alaska Salmon Company.

The salting plant of the Northwestern Packing Company, on Kvichak Bay, was sold to and operated by Nelson, Olsen & Co. The Lagoon Salmon Company operated at Nelsons Lagoon and Mr. Peter M. Nelson on the Igushik River.

continued

JOHN N. COBB COLLECTION, UNIVERSITY OF WASHINGTON LIBRARY

Tonka, Pacific Coast & Norway Packing Company plant on Wrangell Narrows 6 miles south of Petersburg. The cannery moved here about 1906.

HATCHERIES.

Seven salmon hatcheries were operated during the season of 1907-8, as follows:

Name.	Location.	Owners and operators.
Fortmann	Naha Stream	Alaska Packers Association.
Karluk	Karluk River	Do.
Klawak	Klawak Lake	North Pacific Trading and Packing Co.
Hetta	Hetta Lake	Northwestern Fisheries Co.
Quadra	Quadra Stream	Do.
Yes Lake	Yes Lake	U. S. Bureau of Fisheries.
Afognak	Afognak Island	Do.

Source: Marsh and Cobb, *Fisheries of Alaska, 1908,* p. 15.

Right—*Shakan Salmon Company cannery at Shakan, Alaska, circa 1910. Located at head of Shakan Bay east of Hamilton Island, on northwest coast of Kosciusko Island.* Below—*Pacific Coast & Norway Packing Company cannery at Petersburg, early 1900's.*

PHOTOS FROM JOHN N. COBB COLLECTION, UNIVERSITY OF WASHINGTON LIBRARY

From: *Pacific Fisherman, November 1917,* p. 18.

UNION BAY CO. REPORTS GOOD SEASON.

The Union Bay Fisheries company, operating a one-line salmon cannery at Union Bay, Cleveland Peninsula, Alaska, reports a very successful season, making a total pack of over 39,000 cases, which is an increase of 65 per cent over 1916. The company is preparing to make the plant even more efficient for the coming season, though no increase in capacity is proposed. Credit for the season's success is given to N. M. Tate, vice-president and superintendent. L. R. Ellson is president and general manager of the company, and E. J. Wright, secretary. The output is handled by the G. W. Hume company.

REBUILDING EXCURSION INLET CANNERY.

A large shipment of lumber for the reconstruction of the Astoria & Puget Sound Canning company's cannery on Excursion Inlet, Alaska, went north on the steamer Curacao, leaving Seattle, Oct. 25.

In southeast Alaska the run of red salmon was small, but in certain sections this was more than made up by a large run of humpbacks. This year the cannery men in the vicinity of Ketchikan, in order to prevent an expensive fight for red salmon, the species most sought after, met and agreed to so divide up these fish that each plant would get a quantity proportioned to its total outfit. By agreement of the fishermen operating in Boca da Quadra, this year no fishing was allowed inside of a line drawn from the old saltery, just outside the mouth of Quadra Stream, across to the opposite headland. This was done to protect the stream run of red salmon, the old hatchery at the head of Quadra Lake having been reopened.

DISASTERS

So far as loss of life is concerned, this has been the most disastrous year the salmon industry has ever known.

On August 19, as the ship *Lucille*, belonging to the Red Salmon Canning Company, of San Francisco, Cal., was leaving the Ugashik River with the season's pack and cannery crew, she was caught in a gale and driven ashore, where she became a total wreck. The cargo, consisting of about 39,000 cases of canned salmon and 802 barrels and 16 tierces of salted salmon, was entirely destroyed, but fortunately no lives were lost. The *Lucille*, built in Freeport, Me., in 1874, had a gross tonnage of 1,402 and a net tonnage of 1,297.

A disaster without parallel in the history of the Alaska salmon industry, or in that of any of the other fisheries of this country, occurred the morning of September 20, when the Alaska Packers Association bark *Star of Bengal* (Nicholas Wagner, master), while on her way from the association's cannery at Wrangell with the season's pack and the cannery and fishing crews, was wrecked on Coronation Island, in southeast Alaska, and 111 persons (15 whites, including the foreman, bookkeeper, and machinist; 67 Chinese, 26 Japanese, and 3 Filipinos) lost their lives. The vessel and cargo were a total loss, and of the 138 persons aboard but 27 (17 whites, including the master and first mate; 7 Japanese, 2 Chinese, and 1 Filipino) were saved.

The bark left the cannery the day before the wreck in tow of two tugs, which were to continue with her until the open ocean was reached. Shortly after passing out of

continued

Alaska Packers Cannery at Point Highfield, Wrangell, 1911.

COURTESY OF VANCE SUTTER

Above—*In 1927 forklifts were not generally manufactured and salmon cannery operators built their own.* Right—*Early day fish elevator, circa 1913.*

JOHN N. COBB COLLECTION, UNIVERSITY OF WASHINGTON LIBRARY

Sumner Strait a gale sprang up from the south and in a few hours the wind was blowing with terrific force. About 4 o'clock in the morning of the 20th the tugs, finding that despite their exertions the vessel was rapidly drifting onto Coronation Island, cut their tow lines after the bark had anchored, and steamed back to the strait for shelter. The bark held her anchors until 9.30 a.m., when she began to drag. At 10.26 she struck, and in a few minutes was battered to pieces on the rocks. In the meantime a volunteer crew of 5 white men, led by the first mate, had bravely pulled a boat to shore through the raging surf in order to rig up a breeches buoy, but before anything could be accomplished the vessel went ashore. Many of those not carried down with the vessel itself lost their lives in the pounding and grinding mass of wreckage and cargo between the outer rocks and the shore, but few getting through this alive.

The *Star of Bengal* was a bark with a gross tonnage of 1,877 and a net tonnage of 1,694. She was of iron and was built at Belfast, Ireland, in 1873. When launched the vessel was under the British flag and ship-rigged. Later she was transferred to Hawaiian registry, rerigged as a bark, and when the Hawaiian Islands were annexed by this country secured the privilege of flying the American flag.

—Marsh and Cobb, *Fisheries of Alaska, 1908,* pp. 12-15.

Close-up of forklifts made by cannery men.
COURTESY OF VANCE SUTTER

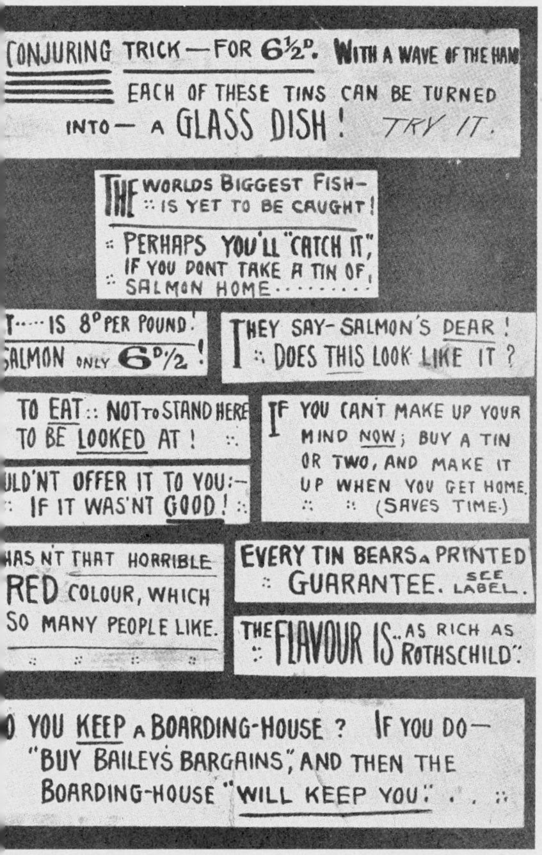

REPRINTED FROM PACIFIC FISHERMAN YEAR BOOK, 1915, P. 79

Steam-driven generators, such as the one at right, provided electric lights in canneries of the early 1900's. Canning lines were driven by steam engines.

From: *Pacific Fisherman, January 1917,* p. 33.

THERMOMETERS FOR CANNERYMEN.

We have recently received from The Foxboro company of Foxboro, Mass., copy of their Bulletin No. 104, illustrating their very complete line of indicating and recording thermometers. This bulletin is handsomely illustrated and describes clearly the actuating principles and the many marked improvements which they have embodied in their design, without a lot of superfluous diction as to their superiority. The catalogue has been arranged so as to make it easy for a prospective customer to pick out a thermometer to meet his requirements, a thing difficult to do from some of the catalogues issued on this subject. Salmon cannerymen will be especially interested in this bulletin as they are rapidly being converted to the belief that an automatic system of temperature control is an absolute necessity in a canning plant.

COURTESY OF WALTER YONKER

Right—*The "Iron Chink" in 1904 was heralded as the first great innovation in the salmon canning process because it eliminated the need for Chinese laborers to do the hand butchering, cleaning and trimming of fish. Eventually the "Iron Chink" was capable of cleaning in excess of 3,000 fish per hour.*

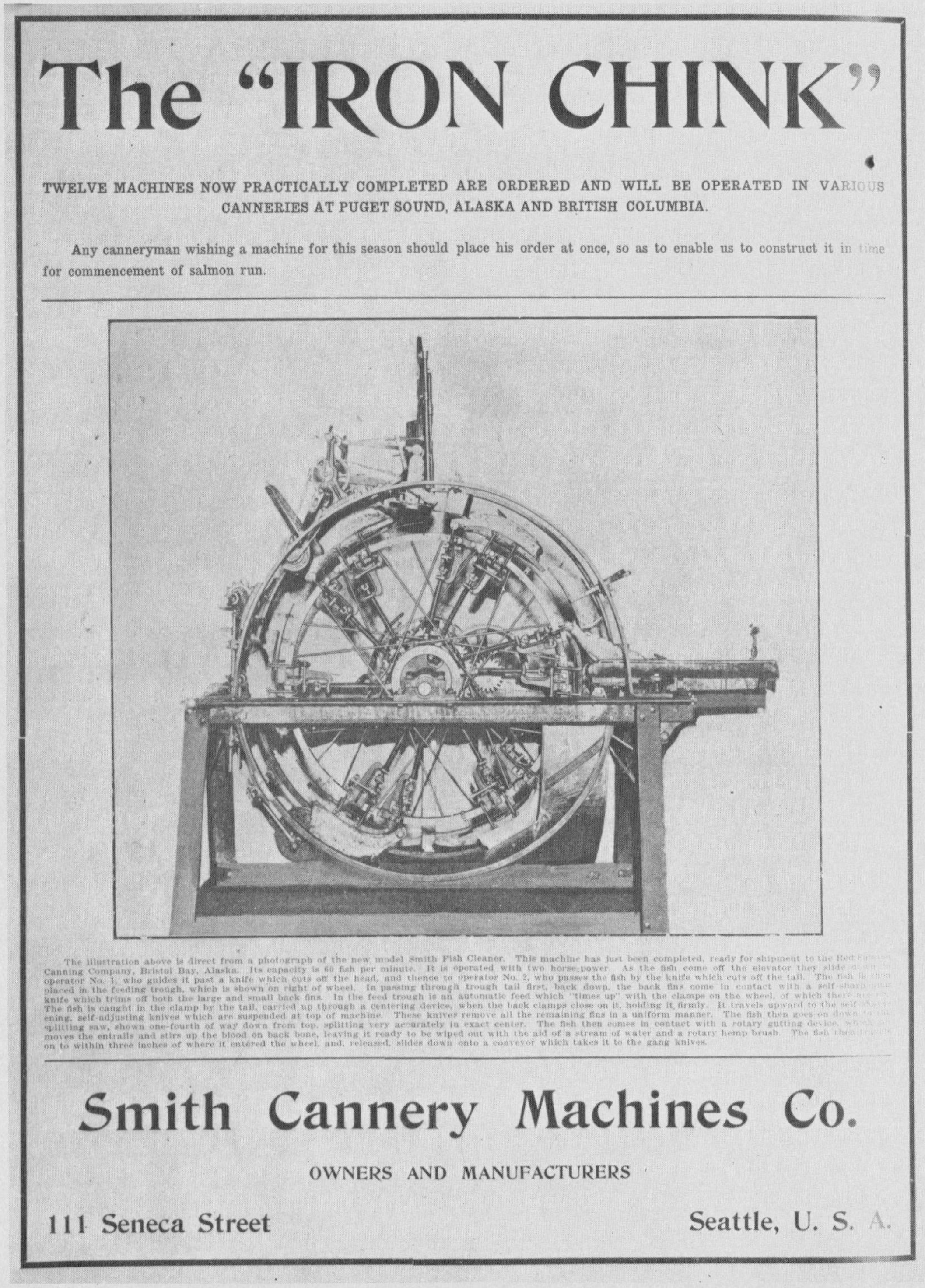

The "IRON CHINK"

TWELVE MACHINES NOW PRACTICALLY COMPLETED ARE ORDERED AND WILL BE OPERATED IN VARIOUS CANNERIES AT PUGET SOUND, ALASKA AND BRITISH COLUMBIA.

Any canneryman wishing a machine for this season should place his order at once, so as to enable us to construct it in time for commencement of salmon run.

The illustration above is direct from a photograph of the new model Smith Fish Cleaner. This machine has just been completed, ready for shipment to the Red [illegible] Canning Company, Bristol Bay, Alaska. Its capacity is 60 fish per minute. It is operated with two horse-power. As the fish come off the elevator they slide down to operator No. 1, who guides it past a knife which cuts off the head, and thence to operator No. 2, who passes the fish by the knife which cuts off the tail. The fish is then placed in the feeding trough, which is shown on right of wheel. In passing through trough tail first, back down, the back fins come in contact with a self-sharpening knife which trims off both the large and small back fins. In the feed trough is an automatic feed which "times up" with the clamps on the wheel, of which there are [illegible] The fish is caught in the clamp by the tail, carried up through a centering device, when the back clamps close on it, holding it firmly. It travels upward to the self-sharpening, self-adjusting knives which are suspended at top of machine. These knives remove all the remaining fins in a uniform manner. The fish then goes on down to the splitting saw, shown one-fourth of way down from top, splitting very accurately in exact center. The fish then comes in contact with a rotary gutting device, which removes the entrails and stirs up the blood on back bone, leaving it ready to be wiped out with the aid of a stream of water and a rotary hemp brush. The fish then travels on to within three inches of where it entered the wheel, and, released, slides down onto a conveyor which takes it to the gang knives.

Smith Cannery Machines Co.

OWNERS AND MANUFACTURERS

111 Seneca Street Seattle, U. S. A.

COURTESY OF NATIONAL FISHERMAN

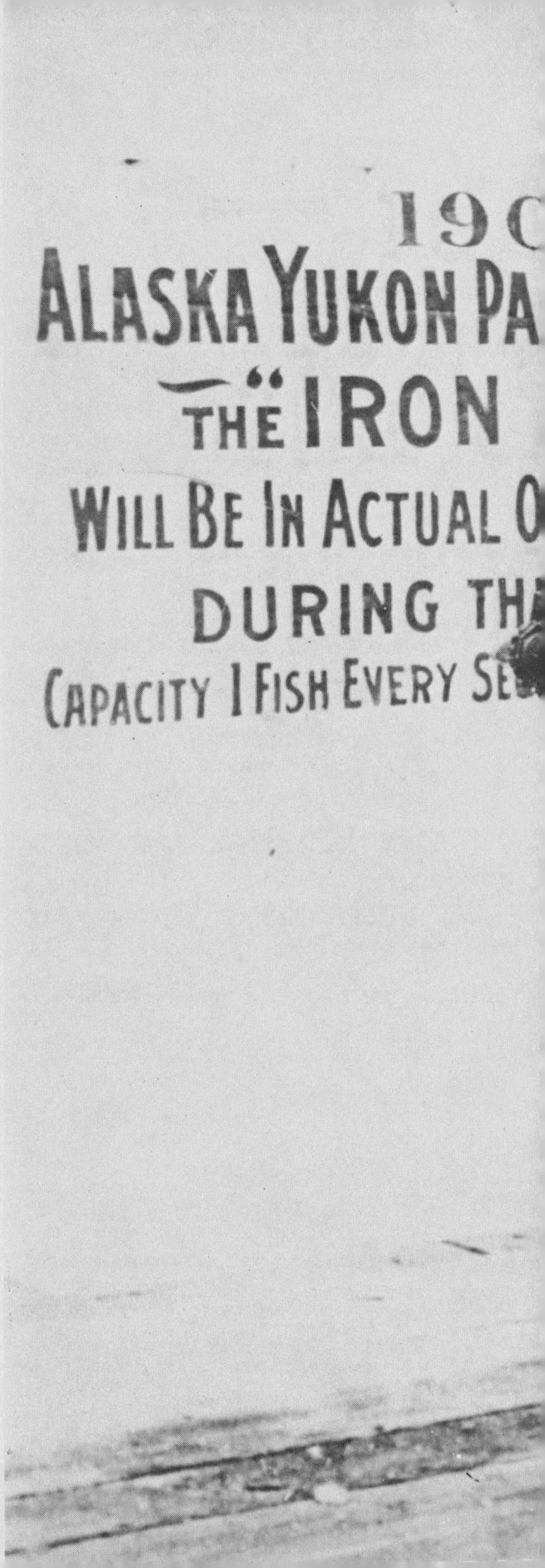

From right at rear are O. A. Smith, inventor of the "Iron Chink" and William Calvert of the San Juan Fishing & Packing Company. Others in photo are unknown.

COURTESY OF JAY GAGE

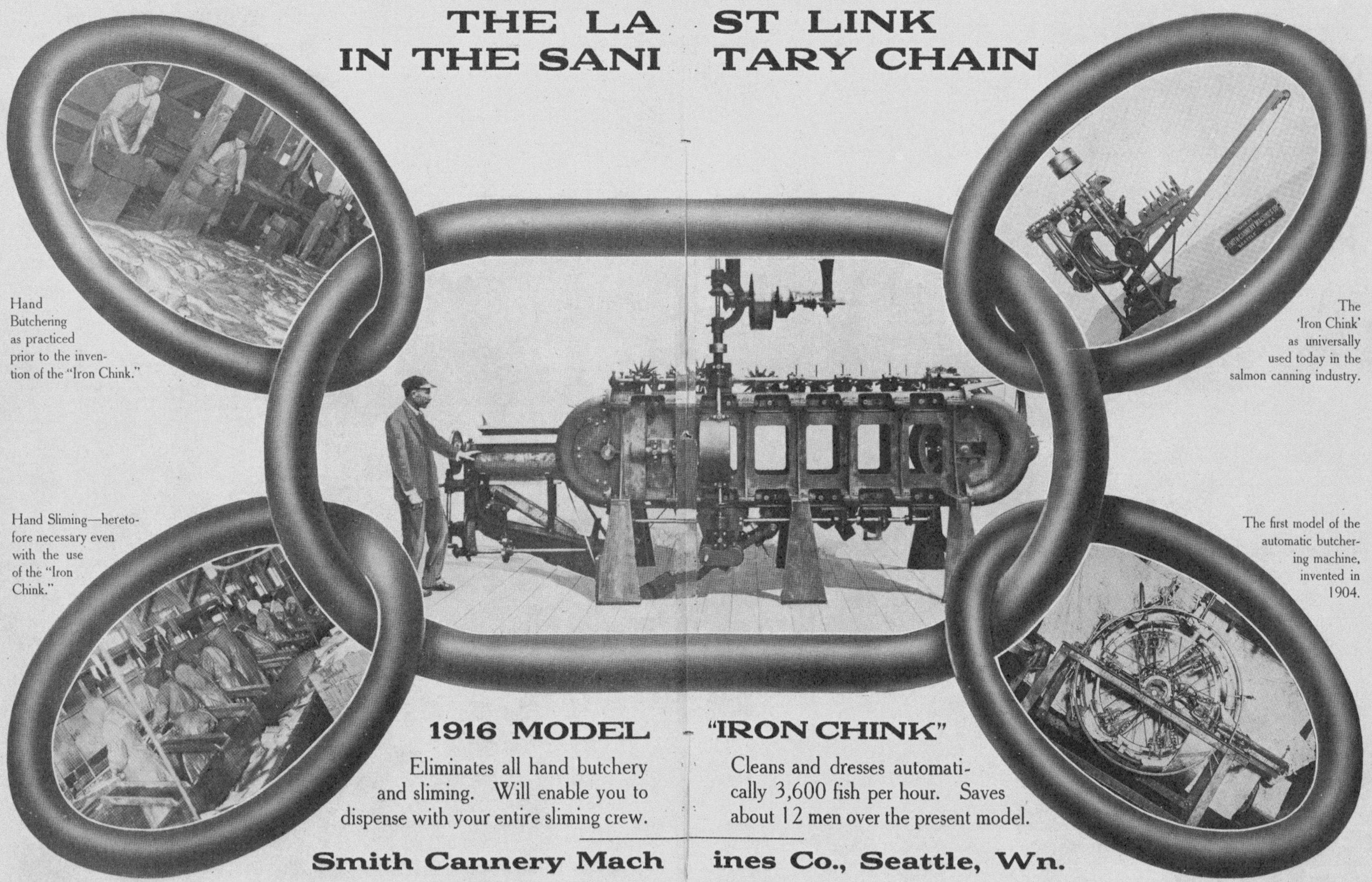

REPRINTED FROM PACIFIC FISHERMAN YEAR BOOK, 1914, PP. 62-63

FRED HEICK COLLECTION, SAN FRANCISCO MARITIME MUSEUM

Above left—*Salmon tins at Alaska Packer's Association cannery, 1916.* Above—*Cans at Chignik cannery.* Left—*Lacquering newly made cans. Raw tin plate was used to make the cans.*

COURTESY OF HAROLD BRINDLE

COURTESY OF JOHN GILBERT

Above—*This Howe soldering machine originally burned coal, but was converted to kerosene by the Astoria Iron Works.*

ADVANCE IN PRICES
FLOUR 1.90
SUGAR 9.00
MEAT 1st 30 2nd 55
CORN
NATIONAL

From: *Pacific Fisherman*, November 1917, p. 15.

New Salmon Canning Projects

Southeastern Alaska.

It is announced that the Northland Trading & Packing company of Tacoma will install a cannery in connection with its salting and mild curing plant at Saginaw Bay, Alaska. A wharf, warehouse and cannery buildings were completed in October, and the equipment will be installed during the winter. Hans Heidner, president of the company, has just returned from a visit to the plants at Saginaw Bay and Port Alexander, leaving M. B. Dahl, the manager, in charge of winter operations.

The Deep Sea Salmon company, which has been operating on Knik Arm, and also at Cape Edwards in the Southeastern Alaska district, is building a new cannery at Port Althorp, off Icy Strait below Idaho Inlet. Most of the buildings have already been completed. A three-line plant will be installed, and one line will probably be moved from the Knik cannery to the new location.

William Semar, for several years manager of the Pacific Mildcure company's operations in Alaska, has resigned that position with the intention of building a cannery at Sitka. Orders for material have been placed and work will be started on the buildings by the first of the year. Associated with Mr. Semar is C. J. Alexander, who has been for some time vice-president of the Hoonah Packing company; and it is understood that several Juneau and Sitka parties are interested.

The Icy Straits Packing company, which this year disposed of its catch to other canneries, has preparations well under way for a new cannery on Idaho Inlet. This company has had considerable success in the operation of traps in the Icy Strait district. Henry Shattuck, of Juneau, the president, has recently been visiting Seattle and San Francisco making arrangements for the new plant. J. L. Freeburn is vice-president, and H. L. Faulkner secretary-treasurer, the largest stockholder being Bart L. Thane.

The Southeastern Alaska Cannery company has been incorporated at Seattle, with a capital stock of $100,000, by Haakon B. Friele, Howard W. Sanders and Edward W. Allen. No details as to the company's plans have yet been announced.

It is reported that T. S. Gorman, who operated a salting plant this year at Sitka Hot Springs, is making preliminary arrangements to put in a canning plant in that vicinity.

The North Pacific Trading & Packing company, which has had a cannery at Klawak for some years, is figuring on putting in a new plant this winter. Hootz Bay and Kelp Bay are mentioned as possible locations for the new cannery.

A. H. Humpheries, who has been looking over the Chatham Strait district, states that he, in partnership with Royal Shepard, will put in a herring packing plant in the vicinity of Killisnoo, and will probably have a cannery to handle salmon, clams, etc.

The Alaska Sanitary Packing company, which has been operating at Wrangell, is making arrangements to put in a new plant at Cape Fanshaw, Frederick Sound.

The Annette Island Packing company's new cannery at Metlakahtla, which could not be finished in time for the 1917 pack, will be brought to completion for the new season's work.

The Fidalgo Island Packing company is considering some further development in Southeastern Alaska, where Otto Sutter is now looking over the ground though as far as can be learned nothing definite has yet been decided.

Plans for the new cannery projected by Capt. A. E. Wadhams, backed by the Columbia Salmon company, in the Tenakee Inlet district, are still in the tentative stage, as are those of the Lisianski Packing company, mentioned in Pacific Fisherman for September.

Other projects, of which more or less uncertain reports have been received, are as follows: Jaeger, Ericson and Peterson, on Lynn Canal; Robert Semple and J. C. McBride, Gedney Harbor; V. A. Paine, Juneau; Alaska Packing & Navigation company, Pavloff Harbor; and W. H. Spaulding, on Chatham Strait.

Aside from these new projects, reconstruction is now under way on the two canneries that were burned this year: the Sunny Point plant near Ketchikan, and the Astoria & Puget Sound cannery at Excursion Inlet.

From: *Pacific Fisherman*, January 1917, p. 16.

MORE CANNERIES CHANGE HANDS.

The year 1916 has witnessed several important changes in cannery ownership, notably the sale of the North Alaska Salmon Co., to Libby, McNeill & Libby, and the Alaska Salmon Co., to C. P. Hale and friends, and the middle of December witnessed the consummation of still another by which the Roe Point, Alaska, cannery of the Lindenberger Packing Co., and the Seldovia, Alaska, plant of the Columbia Salmon Co., (controlled by the Lindenberger interests) were transferred to the Northwestern Fisheries Co.

Under the terms of the sale all of the shore property, together with the vessels, boats, scows, fishing gear, etc., were turned over to the new owners, for a compensation which is unnamed but which is said to have been satisfactory to both parties.

The Roe Point cannery was built in 1911, and is up-to-date in every particular. Owing to the short run in the Ketchikan district during the season of 1916, the pack at this plant was quite a small one—about 36,000 cases—but the pack in 1915, which was more nearly normal, was 81,678 cases.

The Seldovia plant was built by the Seldovia Salmon Co., in 1911, and after a checkered career finally went into the hands of a receiver in 1915. In the fall of the same year it was sold by order of the court, and the new purchasers, the Lindenberger interests, organized the Columbia Salmon Co., for the purpose of operating it. During the season of 1916 the cannery made a good pack, putting up about 37,000 cases.

Opposite—*Naknek Packing Company store at Naknek, Alaska. Note the prices posted on the high shelf at right of photo.*

COURTESY OF WALTER YONKER

Unused salmon cannery at Tanglefoot Bay, near Karluk, 1897.

REPRINTED FROM MOSER, SALMON FISHERIES OF ALASKA, PLATE 56

From: *Pacific Fisherman,* September 1917, p. 25.

DECISION IN KAKE PACKING CO. SUIT.

Creditors of the Kake Trading and Packing company will recover approximately $7500, as the result of a decision handed down by the U. S. Circuit Court, sustaining a former decision in favor of the plaintiffs. The suit was originally brought at Juneau by V. A. Paine as trustee for the Kake Trading and Packing company, on behalf of the creditors, against the Sanborn-Cutting company, which had acquired the Kake property.

From: *Pacific Fisherman,* August 1917, p. 35.

GET RICH BUYING SALMON CANNERY STOCK!

Small contributions are gratefully received by Wm. P. Springer & Son, New York block, Seattle, it would appear from an advertisement placed by them in a recent issue of the Seattle Star, in which a "limited proportion of ownership" is offered to the public in an Alaska salmon canning project, and it is stated that "as little as $100 and not to exceed $250 will be accepted from individuals," and "deferred monthly payments can be arranged for." The parties named, according to the advertisement, "are financial agents of a corporation holding a cannery site and valuable trap locations in Alaska" (no mention being made of the difficulty of "holding" such sites and locations when not actually in use). The ad goes on to say "We can cite instances of very large profits that salmon canneries have made"; and states that its "holdings" are near the canneries named in clippings reproduced from the Seattle "Times" and "Post-Intelligencer" of July 23. These clippings refer to the heavy run at the Hawk Inlet, Hoonah and Funter Bay canneries.

The advertisement was gotten up especially to appeal to the "man on the street," being headed conspicuously "Put in Your Hundred Dollars," and making free use of black faced type, in statements about the fortunes made in salmon canning.

From: *Pacific Fisherman,* May 1917, p. 21.

BRISTOL BAY PACKING CO.

Notwithstanding a short delay on account of cannery materials the vessels of the Bristol Bay Packing Co. all sailed before May 5, when the company's steamer, Costa Rico, left San Francisco Bay. The bark McLaurin, which has been in the general cargo trade all summer, returned last month from Sydney, Australia, with a load of copra in the very good time of 56 days. L. A. Pedersen, head of the company, says he had no special difficulty in getting his cannery crew together, though the company has increased its capacity to more than double that of last year.

SUMMARY OF CANNED SALMON PACK, 1914.

Districts—	Cases	Value
Alaska	4,067,832	$19,719,942
Puget Sound	792,860	5,095,839
Columbia River	455,500	3,707,770
Sacramento River	17,315	157,750
Outside Streams	200,236	1,286,635
British Columbia	1,111,039	7,654,268
Siberia	136,500	999,900
Total	6,781,282	$38,622,104

Source: *Pacific Fisherman Year Book,* 1914, p. 35.

Source: *Pacific Fisherman Year Book*, 1914, p. 38.

1914—SOUTHEAST ALASKA CANNED SALMON PACK—1914.

NAME	CANNERY LOCATION	KINGS 1-lb. Talls	KINGS 1-lb. Flats	KINGS ½-lb. Fl'ts 8 doz.	REDS 1-lb. Talls	REDS 1-lb. Flats	REDS ½-lb. Fl'ts 8 doz.	MEDIUM REDS 1-lb. Talls	MEDIUM REDS 1-lb. Flats	MEDIUM REDS ½-lb. Fl'ts 8 doz.	PINKS 1-lb. Talls	PINKS 1-lb. Flats	PINKS ½-lb. Fl'ts 8 doz.	KETAS, CHUMS 1-lb. Talls	KETAS, CHUMS 1-lb. Flats	KETAS, CHUMS ½-lb. Fl'ts 8 doz.	TOTAL Full Cases
Admiralty Trading Co.	Gambier Bay	Did not operate															
Alaska Packers' Assn.	Loring and Wrangell				10,358			1,214			92,772			42,803			147,147
Alaska Pacific Fisheries	Yes Bay, Chomley and Chilkoot	72			11,742		15,416	2,757		1,775	36,549			75,931			144,242
Alaska Fish Co.	Waterfall	4,132			1,220			1,725			8,885			11,778			27,740
Alaska Sanitary Pkg. Co.	Wrangell	160		57	1,561		1,205	2,000			16,282			3,291			24,556
Astoria & Puget S'd Pkg. Co.	Excursion Inlet				9,086	3,905	1,422	5,156			12,876			12,489			44,934
Barnes, F. C., & Co.	Lake Bay	635			1,757			2,414			17,050			8,893			30,749
Canoe Pass Pkg. Co.	Canoe Pass	Did not operate															
Deep Sea Salmon Co.	Cape Edwards			73	8,950	1,470		2,095			10,000			6,000			28,588
Fidalgo Island Pkg. Co.	Ketchikan						8,087	621		1,069	52,447		1,570	21,409			85,203
George Inlet Pkg. Co.	George Inlet				587			166			7,500			5,088			13,341
Gorman & Co.	Kasaan, Shakan	6	15		733	2,629	1,863	735	83	24	35,142	7,220	533	60,409	5,568	373	115,333
Harris, P. E., & Co.*	Hawk Inlet	258			4,555			2,662			22,302			15,409			45,186
Hidden Inlet Can. Co.	Hidden Inlet				350			300			13,499			8,292			22,441
Hoonah Pkg. Co.	Hoonah				13,583			5,195			17,902			12,634			49,314
Hume, Herbert	Nakat Inlet				315			211			12,375			2,723			15,624
Irving Pkg. Co.	Karheen				4,476			1,781			8,107			5,078			19,442
Kasaan Company	Kasaan	Shown under Gorman & Co.															
Kuiu Island Pkg. Co.	Pt. Beauclerc	148			2,148			241			8,758			8,583			19,878
Lindenberger Packing Co.	Craig				1,575			1,531			14,994			11,200			29,300
Lindenberger Packing Co.	Roe Point				1,863			1,252			32,050			18,000			53,165
Metlakahtla Industrial Co.	Metlakahtla	Did not operate															
Myers, Geo. T., & Co.	Chatham				12,984			4,433			25,210			25,842			68,379
North Pacific Pkg. & Trdg. Co.	Klawack				1,711			1,375			12,557			14,321			29,964
Northwestern Fisheries Co.	Quadra, Hunter Bay, Santa Ana, Dundas Bay	109			10,505	4,718		5,230			70,939	1,928		50,174			143,603
Pacific American Fisheries	Excursion Inlet		460		23,300	9,100	2,450	11,200	200		26,000			18,600			91,310
Pac. Coast & Norway Pkg. Co.	Petersburg			20	221	4,092		3,520			10,432	117		8,701			27,103
Pillar Bay Pkg. Co.	Point Ellis				3,095		395	963			6,700			7,000			18,153
Point Warde Pkg. Co.	Point Warde				1,182			164			19,617			19,701			40,664
Pure Food Fish Co.	Ketchikan				4,429			691			19,078			5,518			29,716
Shakan Salmon Co.	Shakan	Shown under Gorman & Co.															
Sanborn-Cutting Co.	Kake	1,505	340	983	204			930			7,098			11,038			22,098
Sanborn-Cram Co.	Burnett Inlet	1,409			800			540			21,050			10,100			33,899
Starr-Collinson Pkg. Co.	North Arm				1,913			298			6,026			6,944			15,151
Sunny Point Pkg. Co.	Sunny Point	18			681			278			11,212			14,044			26,233
Swift-Arthur-Crosby Co.	Heceta Island	1,144			1,310			1,636			1,678			1,203			6,971
Skowl Arm Pkg. Co.	Skowl Arm	Did not operate															
St. Elias Pkg. Co.	Alsek	Did not operate															
Taku Canning & Cold Stg. Co.	Taku Harbor			166	3,025		3,378	10,177		1,712	22,496			16,671			57,625
Tee Harbor Pkg. Co.	Tee Harbor				9,847			4,968			12,791			10,385			37,991
Thlinket Pkg. Co.	Funter Bay	357			31,164	3,289	2,521	14,165			38,650			29,064			119,210
Wards Cove Pkg. Co.	Ward's Cove				628			304			11,378			2,254			14,564
Wiese Packing Co.	Rose Inlet	20			3,315			593			18,359			9,924			32,211
Yakutat & Southern Rwy.	Yakutat				36,620	7,199		13,779			375						57,973
Total		9,973	815	1,299	221,703	36,402	36,737	107,300	283	4,580	761,136	9,265	2,103	591,494	5,568	373	1,789,031

* Formerly Hawk Fish Co.

1914—CENTRAL ALASKA CANNED SALMON PACK—1914.

NAME	CANNERY LOCATION	KINGS 1 lb. Talls	KINGS 1 lb. Flats	REDS 1 lb. Talls	REDS 1-lb. Flats	MEDIUM REDS 1 lb. Talls	PINKS 1 lb. Talls	KETAS, CHUMS 1 lb. Talls	TOTAL Full Cases
Alaska Packers Association	Kasilof, Larsen's Bay, Chignik and Alitak	3,223		113,969		12,966	90,844	2,414	223,416
Columbia River Packers Association	Chignik	63		34,248		1,094	1,255	1,228	37,888
Fidalgo Island Packing Company	Point Graham	210	218	9,761	12,340	2,827	14,749	2,423	42,528
Kadiak Fisheries Company	Kodiak	43		13,142		407	5,178	57	18,827
Libby, McNeil & Libby	Kenai	4,315		25,779		3,912	10,368	54	44,428
Northwestern Fisheries Company	Orca, Kenai, Uyak, Chignik	2,004		131,764		11,227	45,183	1,217	191,395
Pacific American Fisheries	King Cove	80		47,200		590	11,800	10,300	69,970
Seldovia Salmon Company	Seldovia	976		6,907		3,716	16,453		28,052
Total		10,914	218	382,770	12,340	36,739	195,830	17,693	656,504

1914—WESTERN ALASKA, OR BRISTOL BAY, CANNED SALMON PACK—1914.

NAME	CANNERY LOCATION	KINGS 1-lb. Talls	KINGS 1-lb. Flats	KINGS ½-lb. Flats 8 doz.	REDS 1-lb. Talls	REDS 1-lb. Flats 8 doz.	REDS ½-lb. Flats	MEDIUM REDS 1 lb. Talls	PINKS 1 lb. Talls	KETAS OR CHUMS 1 lb. Talls	TOTAL Full Cases
Alaska Packers Association	Nushagak (2), Koggiung (2) Naknek (3) and Ugaguk	5,629			760,287			7,193	23,167	20,616	816,892
Alaska-Portland Packers Assn.	Nushagak	3,746			72,967		2,948	130	532	6,900	87,223
Alaska Salmon Co.	Wood River	1,644			34,117			19		3,411	39,191
Bristol Bay Packing Co.	Kvichak				45,000						45,000
Columbia River Packers Assn.	Nushagak	3,597			52,473			127		6,222	62,419
Libby, McNeill & Libby*	Nushagak and Koggiung			2,107	88,435		1,571			3,599	95,712
Midnight Sun Pkg. Co.	Kotzebue Sound							900			900
Naknek Packing Co.	Naknek	210			99,088	337	221		387		100,243
North Alaska Salmon Co.	Koggiung, Lockonock, Nushagak and Ugaguk	1,721			175,295	3,373	2,389	128	3,613	3,272	189,791
Northwestern Fisheries Co.	Nushagak	4,380			44,569			393		3,330	52,672
Pacific American Fisheries	Port Moller	700	1,000	25	75,400	5,300	4,750				87,175
Red Salmon Canning Co.	Ugashik	187			43,102				1,790		45,079
Total		21,814	1,000	2,132	1,490,733	9,010	11,879	8,890	29,489	47,350	1,622,297

* Formerly Alaska Fishermen's Pkg. Co.

From: *Pacific Fisherman*, July 1917, p. 35.

Preserving Fish Without Ice

Opportunity for the Small Producer

"SHERMAN'S" Fish Sterlizing Co., Ltd., 1416 Standard Bank Bldg., Vancouver, B. C., is the owner of the Henderson Process for preserving fish without ice.

A. H. Sherman, proprietor of the Defiance Packing Co., is the Britisher who introduced this process into Canada, being impressed with its importance from the fact that the Board of Agriculture and Fisheries, 43 Parliament Street, London, England, reported in February, 1917, that "there appears to be no ground for doubting Mr. Henderson's claims as to the practicability of the process on a commercial scale." When a British government board puts its seal of approval on a fish preserving process, it is good enough for all Britishers, is Mr. Sherman's opinion.

Many tests with an experimental plant have been carried out in Vancouver, and their results have been satisfactory to all concerned. After a recent test, Mr. Sherman received the following unsolicited testimonial from Edward G. Taylor, inspector of Dominion Fisheries, Nanaimo, B. C.

"The test proved entirely successful and the process was exceedingly simple, and no ice was used at any time during the process. I also ate salmon which had gone through this process, and had been out of the water for fourteen days, and this salmon was just as fresh and firm as if it had been taken out of the water that very day. The bone of the salmon was strong, sound and sweet, proving conclusively that the process was entirely successful in preserving fish without ice for at least fourteen days, and from what I have seen, I have no doubt it would keep a very much longer time. I believe the process will be a great boon to the country and be an immense factor in the development of the fishing industry."

Some of the processed fish was expressed to W. A. Found, Department of Naval Service, Dominion Fisheries, Ottawa, and he wrote after sampling it: "There was nothing in the taste of either fish that suggested to me that they had not been cured immediately following their being landed."

It is claimed by the owners of this patented process for preserving fish without ice, that because no ice is needed the cost of preservation is lessened, and therefore the fish can be sold cheaper.

All the original flavor of the fish is maintained. The process applies to fresh and smoked fish, and also to meats, with which successful tests have already been made.

The whole process from start to finish takes only three hours. Salt, low temperatures and sterilization are the salient features of the process. First the fish is put into a cooling tank filled with water and brought to a low temperature. In half an hour the latent heat in the fish is extracted gradually and entirely. Then the fish is put into a second tank of sea water or fresh water strengthened by the addition of salt. In order to prevent freezing the water is agitated by a pump which draws it off through one pipe and drives it back again through another, passing through a filtering chamber charged with willow charcoal to kill the germs. The extremely low temperature of the salt solution seals up the pores of the fish and prevents saturation, acting as an antiseptic protection on the outside. After three hours the fish is taken out and presents a fresh appearance. It is impervious to decay for ten days, and may be kept in a cool room for months.

From: *Pacific Fisherman Year Book*, 1914, p. 75.

Fish Coming Into Its Own

The diminishing supply of land animals capable of supplying human food, and the consequent increase in the cost of same, in many instances reaching such an altitude as to preclude their regular use by people of moderate or poor circumstances, has caused the attention of the economist and scientist to be directed to fish as an excellent substitute, one that is also cheaper, more nutritious and wholesome than meat.

As ordinarily employed the term fish includes, besides the fish proper, many other water animals, as oysters, clams, and other mollusks; lobsters, crawfish, crabs and shrimps, and turtle and terrapin.

But few people have any conception of the present vast extent of our salt and fresh water fisheries, and yet, should the demand warrant it, this output could be doubled and trebled. Complete and accurate figures of the annual catch of the commercial fishermen of the United States are not available, but the best estimate, based upon accurate data for certain sections, and estimates from the best sources possible covering the other regions, is to the effect that about 2,400,000,000 pounds of fishery products was taken from our waters last year. This enormous catch had a value to the fishermen of approximately $70,000,000. The canning, salting, and otherwise preserving of a considerable part of this catch gave employment to an army of additional employees and added from 30 to 50 per cent to the original value of the catch.

The population of the United States at the last census was 91,972,266, and if this enormous catch were to be equally divided each man, woman and child in the country would have consumed approximately 26.1 pounds of fish in 1914. This is a very small percentage, being only about one-half pound per week, and if it were to be trebled it would mean a vast increase in the number of fishermen employed and in the amount of business done by the dealers in fish and the manufacturers of fishing gear, vessels, boats, etc., thus giving employment to more than twice the number now employed.

FISH.	Per cent protein.	Cost per pound, cents.
Haddock	13	7
Herring	19	8
Bluefish	19	10
Codfish	18	12
Eels	18	15
BEEF.		
Chuck	19	24
Rump	19	24
Round	21	20
Sirloin Steak	19	30
Ribs	18	26

REPRINTED FROM MOSER, SALMON FISHERIES OF ALASKA, PLATE 58

REPRINTED FROM MOSER, SALMON FISHERIES OF ALASKA, PLATE 57

Left—*Cannery at Uyak Bay on northwestern side of Kadiak (Kodiak) Island, 1897. This cannery is not identified, but at this time there were two canneries—one operated by Pacific Steam Whaling Company, the other by Hume Brothers & Hume.* Below—*Redfish hatchery at Karluk was built in 1896 and operated by the Alaska Packers' Association. The hatchery was built on the southern bank of the Karluk River, Kodiak Island, about 2 miles from the outlet. The first season, fall of 1896 to spring of 1897, 2,500,000 red salmon were hatched and liberated.*

From: *Pacific Fisherman,* September 1917, p. 25.

FEDERAL CHILD LABOR LAW.

While child labor has not been extensively employed in the Pacific fishing and fish canning industries, those interested may avoid trouble by taking notice of the new Federal child labor law, which took effect Sept. 1. The new law, applying to the manufacture of all goods going into interstate commerce, forbids the employment of children under 16 years of age in mines and quarries, and those under 14 years of age in mills, canneries, manufacturing establishments, etc. It also provides that no child between 14 and 16 years of age be employed more than eight hours a day or more than six days a week.

From: *Pacific Fisherman,* March 1917, p. 16.

SANITARY CANS FOR NAKNEK CO.

The Naknek Packing Co.'s plant is changing this year from the old style can to the sanitary can, which will involve some changes in the crew. Not quite as many men will be needed as formerly, and the contract work will be done by Japanese. It is thought that about 155 men will be required. The barks Pactolus and B. P. Cheney, which have been anchored in the Oakland estuary all winter undergoing repairs to the rigging and interior, will be drydocked in a few days and are expected to go to San Francisco by the end of March to complete outfitting for the northern stations.

NAVAL ARCHITECTS

Fishing Boats a Specialty

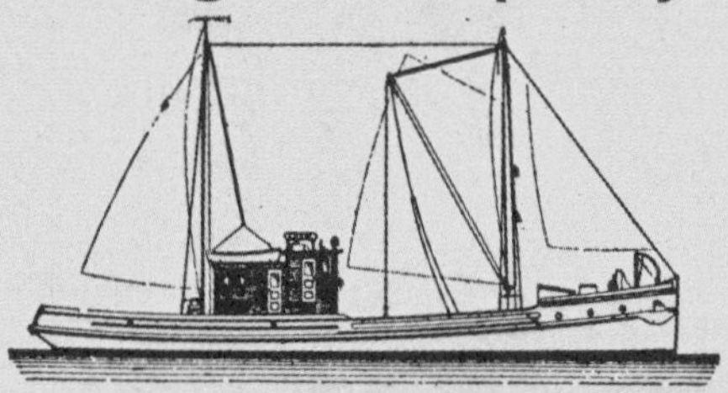

LEE & BRINTON

COLMAN DOCK, SEATTLE **Main 4879**

Established 1867.

THE VILTER MFG. CO.

1034 Clinton St., MILWAUKEE, WIS.

ICE and REFRIGERATING MACHINES Vertical or Horizontal

Horizontal Double-Acting Refrigerating Machine

AMMONIA FITTINGS,
AMMONIA CONDENSERS,
SHARP-FREEZER COILS,
DOUBLE PIPE BRINE COOLERS,
ETC.

Branch Offices in all Large Cities.

Write for Bulletins

REPRINTED FROM PACIFIC FISHERMAN YEAR BOOK, 1914, P. 78

Phillips Sheet & Tin Plate Co.

WEIRTON, WEST VA.

Actually the Largest Independent Manufacturers of Tin Plate in America

---

4,500,000

BASE BOXES

ANNUAL TIN PLATE PRODUCTION

Four Separate Plants

Producing Open Hearth and Bessemer Plate of a Grade Eminently Suitable for Making Cans—Sanitary or Any Other Type

---

The Highest Grade of Coke Tin Plate

ADDRESS OUR AGENTS:

Balfour, Guthrie and Company

OFFICES AT

San Francisco and Los Angeles, Cal. Seattle and Tacoma, Wash.
Portland, Oregon Vancouver, B. C.

REPRINTED FROM PACIFIC FISHERMAN, FEBRUARY 1917, P. 4

REPRINTED FROM PACIFIC FISHERMAN YEAR BOOK, 1917, P. 67

This case contains 330 1-lb. tall salmon can bodies collapsed for shipment

COLLAPSED CANS

Another forward stride for the patrons of the

American Can Company

Seattle San Francisco Portland Los Angeles Bellingham Vancouver, B. C.

The same case holds 48 only 1-lb. tall salmon cans shipped in the ordinary fashion

Collapsing and Reforming Process and Machinery Patented

The Proportion of the Total Salmon Pack Sealed in

SANITARY CANS

IS INCREASING EVERY YEAR

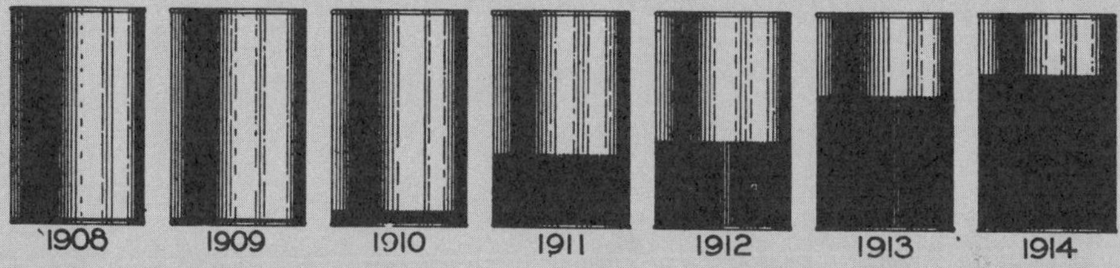

NOTE.—The darkened sectio i of the above cans shows the portion of each year's pack marketed in Sanitary Cans.

Yesterday---

Eight years ago the sanitary can was unknown in the salmon canning industry.

Today---

The sanitary can is the standard of the industry. Two-thirds of the entire world output of canned salmon for 1914 appeared in the new container.

Tomorrow---

The packer who still clings to the soldered can will be as extinct as the "Dodo" if the present rate of progress continues.

American Can Company

SAN FRANCISCO PORTLAND
VANCOUVER BELLINGHAM

REPRINTED FROM PACIFIC FISHERMAN YEAR BOOK, 1914, P. 51

Shipping tin plate to Alaska and making cans in the early spring was discontinued with the sanitary can. The body was soldered, collapsed and 10 cases were shipped in one case. Tops and bottoms were attached during the canning operation.

Pacific Coast Salmon Canner's Seattle=Astoria Excursi

An American Can Company demonstration of the sanitary can drew this group of Pacific Coast salmon packers.

REPRINTED FROM PACIFIC FISHERMAN, NOVEMBER 1912, SUPPLEMENT

Supplement to Pacific Fisherman, November, 1912.

The above group shows the salmon canners who attended the Astoria Excursion as guests of the American Can Co. Those shown are: (1) August Buschman
Co.; (3) John McGowan, P. J. McGowan & Sons; (4) H. Bell-Irving, Jr., Anglo, B. C., Pkg. Co.; (5) J. J. McBain, Anglo, B. C., Packing Co.; (6) O. M. Ash, American
tendent Wiese Pkg. Co.; (8) J. L. Smiley, J. L. Smiley & Co.; (9) D. W. Branch, Libby, McNeill & Libby; (10) R. E. Small, Kelly-Clarke Co.; (11) Wm. Stewart, G
Co.; (13) H. H. Snow, Superintendent Ainsworth & Dunn; (14) Carl A. Sutter, Fidalgo Island Pkg. Co.; (15) J. Rowan, B. C. Packers' Association; (16) George T
Can Co., Portland; (18) F. A. Seuffert, Seuffert Bros.; (19) F. C. Johnstone, Alaska Fish Co.; (20) John L. Myers, Willapa Harbor Fish Co.; (21) P. E. Harris, Haw
F. M. Warren, Jr., Alaska-Portland Packers' Association; (24) C. H. Buschmann, Northwestern Fisheries; (25) E. Shelley Morgan, Schmidt Lithograph Co., Portland; (2
penter, Canoe Pass Pkg. Co.; (28) W. F. McGregor, Alaska Fishermens' Pkg. Co.; (29) Fred Barker, Superintendent Thlinket Pkg. Co.; (30) H. S. McGowan, P. J. McGov
H. Bell-Irving, Sr., Anglo, B. C., Pkg. Co.; (33) C. H. McGowan, P. J. McGowan & Sons; (34) F. C. Barnes, F. C. Barnes & Co.; (35) E. E. Fleming, Hoonah Pkg. Co.; (3
Carlisle Pkg. Co.; (38) K. J. Saeter, Admiralty Trading Co.; (39) J. G. Cotton, Taku Can. & C. S. Co.; (40) Geo. W. Sanborn, Sanborn-Cutting Co.; (41) F. P. Kendall, A
bor Fish Co.; (43) H. W. Phelps, General Manager American Can Co.; (44) Lee H. Wakefield, Apex Fish Co.; (45) B. Lindenberger, Lindenberger Pkg. Co.; (46) Daniel

, Oct. 19-20, 1912

Sea Salmon Co.; (2) E. A. Sims, Hoonah Packing
d Tinplate Co., Portland; (7) E. Larsen, Superin-
vers & Co.; (12) W. E. Persell, Friday Harbor Pkg.
niralty Trading Co.; (17) F. W. Prael, American
Co.; (22) John E. Rice, Lummi Bay Pkg. Co.; (23)
Bigger, American Can Co., Portland; (27) D. C. Car-
ons; (31) R. A. Welsh, Bellingham Canning Co.; (32)
I. Barker, B. C. Packers' Assn.; (37) P. J. Andrae,
Can Co., Portland; (42) Geo. Foster, Willapa Har-
ell, Astoria & Puget Sound Pkg. Co.; (47) Herbert

REPRINTED FROM PACIFIC FISHERMAN YEAR BOOK, 1917, P. 94

REPRINTED FROM PACIFIC FISHERMAN YEAR BOOK, 1917, P. 50

THE FOLLOWING SALMON PACKERS

WILL USE SANITARY SALMON CANS SEALED WITH JOHNSON CLOSING MACHINE IN 1916

Ainsworth & Dunn
Alaska Fish Co.
Alaska Fishermen's Packing Co. (2 canneries)
Alaska Pacific Fisheries (3 canneries)
Alaska Packers Association (17 canneries)
Alaska-Portland Packers Association
Alaska Sanitary Packing Company
Altoona Packing Co.
Anglo-B. C. Packing Co. (8 canneries)
Apex Fish Co.
Astoria & Puget Sound Canning Co. (2 canneries)
Atlin Construction Co.
Barbey Fish Co.
F. C. Barnes Co. (4 canneries)
Bay City Storage & Fisheries Co.
Bellingham Canning Co.
Blaine Packing Co.
British Columbia Canning Co.
British Columbia Packers Association (12 canneries)
Bugge Canning Co.
Canadian Fish & Cold Storage Co.
Canoe Pass Packing Co.
Carlisle Packing Co.
Chetlo Harbor Packing Co.
John F. Childs
Chinook Packing Co.
Clayoquot Sound Canning Co., Ltd
Columbia River Packers Association (6 canneries)
Coquille River Fishermen's Co-op. Packing Co.
Defiance Packing Co., Ltd.
Del Norte Salmon Canning Co.
M. Des Brissay & Co. (2 canneries)
Doyhof Fish Products Co.
Draney Fisheries, Ltd.
Eagle Harbor Packing Co., Ltd.
S. Elmore & Co. (8 canneries)
Everett Packing Co.
Fidalgo Island Packing Co. (4 canneries)
Florence Canning Co.
Friday Harbor Packing Co.
George & Barker Salmon Packing Co.
George Inlet Packing Co.
Glen Rose Canning Co.,Ltd.
Goletas Fish Co.
Gosse-Millard Packing Co., Ltd.
The Graham Co.
Great West Packing Co.
Grooshetsky & Co.
David Hansen Packing Co.
P. E. Harris & Co.
Hidden Inlet Canning Co. (2 canneries)
Hodgson & Graham Co.
Hoquiam Packing Co.
G. W. Hume Co.
Ichigumi & Co.
Jervis Inlet Canning Co.
Kadiak Fisheries Co.
Key City Packing Co.
Kildala Packing Co., Ltd.
Kincolith Packing Co., Ltd.
Klamath River Canning Co.
W. W. Kurtz
Wm. Kyle & Sons
A. H. Lawrence
Libby, McNeill & Libby
Lindenberger Packing Co. (2 canneries)
Lummi Bay Packing Co.
Macleay Estate Co.
P. J. McGowan & Sons (2 canneries)
J. G. Megler & Co.
Midnight Sun Packing Co.
Morse & Son
Naknek Packing Company
Nanaimo Canning Co.
National Independent Fisheries Co.
North Alaska Salmon Co.
North Pacific Trading & Packing Co.
Northern Canning Co.
Northwestern Fisheries Co. (3 canneries)
Oregon Fisheries Co.
Pacific American Fisheries (5 canneries)
Pacific Fisheries & Packing Co.
Pillar Rock Packing Co.
Preston Packing Co., Ltd.
Pure Food Fish Co.
Quathiaski Canning Co., Ltd.
Red Salmon Canning Co.
Sanborn-Cram Company
Sanborn-Cutting Company (2 canneries)
San Juan Canning Co.
San Juan Fishing & Packing Co.
Scottish-Canadian Canning Co.
B. A. Seaborg Co., Inc.
Seattle Packing Co.
Seufert Bros. Co.
Shaw Island Canning Co.
E. A. Sims
Sinclair Island Canning Co.
Skeena River Commercial Co., Ltd.
J. L. Smiley & Co. (2 canneries)
St. Mungo Canning Co.
Starr-Collinson Packing Co.
Steveston Canning Co., Ltd.
Straits Packing Co.
Sunny Point Packing Co.
Surf Packing Co.
Swift-Arthur-Crosby Co.
W. E. Tallant
Tallant-Grant Packing Co.
Thlinket Packing Co.
Tillamook Ice & Cold Storage Co.
Tsutsumi & Co.
Umpqua Co-Operative Packing Co.
John Wallace
Wallace Fisheries, Ltd.
Ward's Cove Packing Co.
Warren Packing Co. (2 canneries)
Wm. Westbrook
Western Canners, Ltd.
Wiese Packing Co.
Willapa Harbor Fish Co.

AMERICAN CAN COMPANY

San Francisco, Portland, Vancouver, B. C., Bellingham

AMERICAN CAN COMPANY

San Francisco, Portland, Vancouver, B. C., Bellingham

REPRINTED FROM PACIFIC FISHERMAN YEAR BOOK, 1917, PP. 52-53

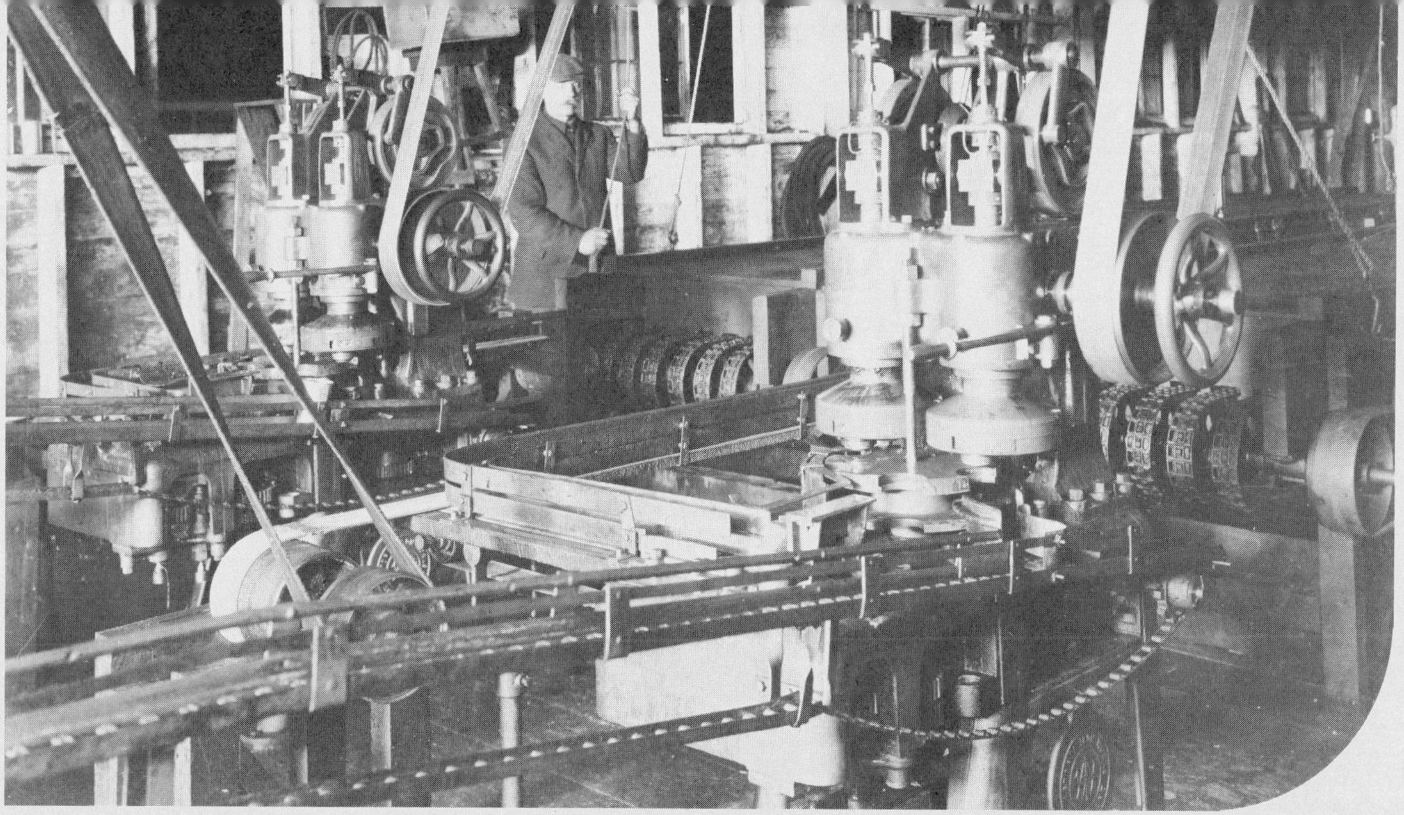

After leaving the exhaust box, where steam created a vacuum in the partially sealed cans, the cans entered the double seamer for final sealing before cooking.

PHOTOS COURTESY OF WALTER YONKER

Early method of loading metal coolers was all muscle. The cans were loaded by hand into metal trays (coolers), then the loaded trays were put into retorts for cooking by steam. When cooking was completed, trays were removed from retorts and cans remained in trays to cool and to await stacking or casing the next day.

From: *Pacific Fisherman Year Book*, 1917, p. 81.

Condensed Facts About Canned Salmon

GRADES OF CANNED SALMON.

The Red, Blueback, Sockeye or Quinault, a small fish ranging from 4 to 12 pounds, with average about 5 pounds; flesh firm and intensely red in color.

The Quinnat, Chinook, Tyee, King or Spring salmon, the largest of all, averaging about 22 pounds, but often much heavier; flesh from pale to deep pink in color and rich in flavor.

The Coho, Medium Red or Silverside, averaging about 6 pounds in weight; flesh firm but much paler in color than the Red salmon.

The Humpback or Pink salmon, the smallest of the five, averaging about 4 pounds; flesh is pale pink and soft.

The Chum or Keta salmon, averaging about 8 pounds; flesh yellowish in color.

The Steelhead, although commercially classed as salmon, properly belongs to the trouts. As they are much sought after by the fresh fish dealers but few are canned.

MARKETS IN WHICH THE DIFFERENT SPECIES ARE CONSUMED.

The Quinnat, Tyee, Chinook or King salmon, principally the United States.

The Red, Blueback or Sockeye, throughout the world.

The Coho, Medium Red or Silverside, principally the United States, although of recent years it has been selling in England.

The Humpback or Pink, principally the United States; some in Australia, South and Central America and the Orient; also of recent years to England and Italy in considerable numbers.

The Chum or Keta salmon, United States; also in 1917 a considerable quantity was sold to Italy.

CANNING THE SALMON.

So far as possible the handling of the fresh salmon is done by machinery. Flat cans are filled by hand, the workers being supplied with gloves for this purpose twice a day.

All grades of salmon are packed in exactly the same manner.

Salmon are canned in three styles of cans, one-pound talls, one-pound flats and one-half pound flats. The shape of the can does not signify any particular grade or quality of fish.

The so-called "Sanitary" cans have no vent holes on the top. The old style process, by which a considerable quantity is still being packed, shows two soldered holes in the top. The first was for the purpose of permitting the steam and gas to escape while passing through the soldering machine, and the second was caused by the cans being vented at the end of the first cooking to permit the surplus gas, steam and water to escape. Both processes turn out an excellent product, although the former is rapidly superseding the latter, mainly for mechanical reasons.

Absolutely no foreign substance, except one-fourth ounce of fine salt to give the fish a flavor, is placed in the can.

FOOD VALUE OF CANNED SALMON.

Prof. W. O. Atwater, the late eminent chemist of the U. S. department of agriculture, whose investigations of the food value of certain products has made his name a household word throughout the world, reports the following as the relative food values of Canned Salmon as compared with other foods:

CANNED SALMON	21.8
Sirloin Steak	16.5
Sugar-cured Ham	14.2
Macaroni	13.4
Eggs	13.1
Spring Chicken	12.0
White Bread	09.0

"The cheapest food is that which supplies nutriment at the lowest cost. One pound of canned Red salmon of the best quality will cost about 16 cents.* The same quantity of bone, muscle, blood and brain building material and body fuels in other foods would cost—

	Cents.
Eggs, strictly fresh (at 34 cents per dozen)	36
Steak, sirloin (at 27½ cents per pound)	33
Mutton, leg (at 19 cents per pound)	32
Chicken, average (at 25 cents per pound)	21½
Ham, smoked (at 18½ cents per pound)	13¼
Pink salmon, canned (at 9 cents per can)	12½

From: *Pacific Fisherman Year Book*, 1916, p. 48.

FISHERY TAXES AND LICENSES IN ALASKA.

Federal Government.

Canned salmon, 4c per case.

Pickled salmon, 10c per barrel.

Salt salmon in bulk, 5c per 100 pounds.

Fish oil, 10c per barrel.

Fertilizer, 20c per ton.

Applications for federal licenses should be made to the Clerk of the Court of the judicial district in which the applicant desires to engage in fishing. Juneau is the headquarters for the Southeast Alaska section and Valdez for the Central and Western sections. Such application should be made before engaging in business, and the applicant will be permitted to operate without the payment of any money until the end of the fishing season, when he will be required to make return of the work done and pay in the money to cover same.

Territorial Government.

Canned King and Red salmon, 4c per case.

Canned Medium Red or Coho salmon, 2c per case.

Canned Pink and Chum salmon, 1c per case.

Salted fish (except herring), 2½c per 100 pounds.

Fish traps, fixed or floating, $100 per annum.

Gill nets (per 100 fathoms or fraction), $1.

Cold storage fish plants—

Doing business of $100,000 or over, $500 per annum.

Doing business of $75,000 or less than $100,000, $375 per annum.

Doing business of $50,000 and less than $75,000, $250 per annum.

Doing business of $25,000 and less than $50,000, $125 per annum.

Doing business of $10,000 or less than $25,000, $50 per annum.

Doing business of $4,000 and less than $10,000, $25 per annum.

Doing business of under $4,000, $10 per annum.

(The "Annual Business" under this section shall be considered the amount paid per annum for the product).

PACIFIC FISHERMAN | PACIFIC FISHERMAN

This Is the Most Complete Line of Salmon Canning Machinery Manufactured by One Company

FILLING MACHINE
(Patented Sept. 24th, 1912.)

The machines we manufacture are for the three sizes of salmon cans and are as follows:
Elevator Fish Cutter.
Filling Machines with Salters attached.
Weighing Machines.
Clinchers.
Endless Diamond Chain Exhaust Boxes.
Double Seamers.
Retorts.
Retort Cars and Coolers.

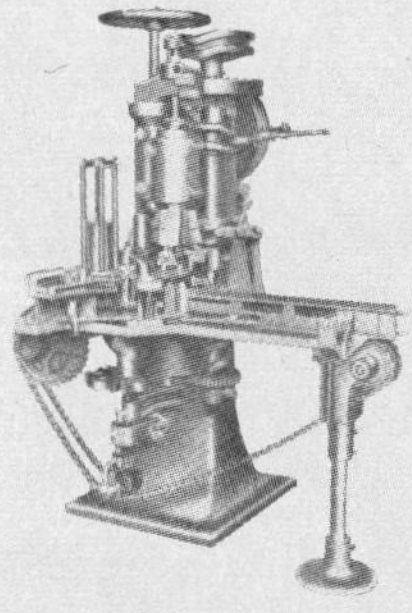

CLINCHER
(Patent applied for.)

This Double Seamer has proven its reliability, having had one successful season's run. Watch for the results of the Clincher and Double Seamer this coming season.

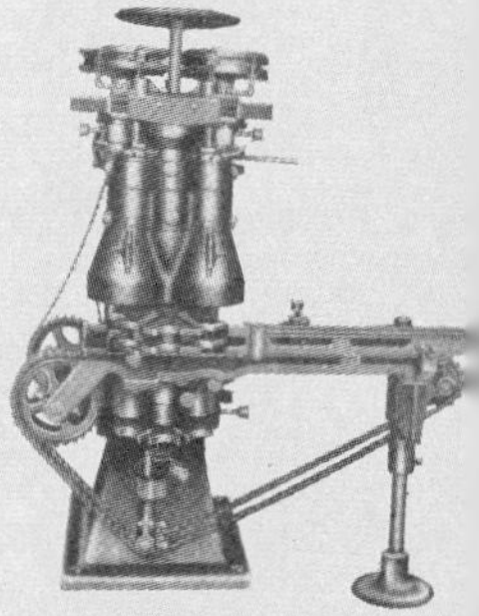

DOUBLE SEAMER
(Patent applied for.)

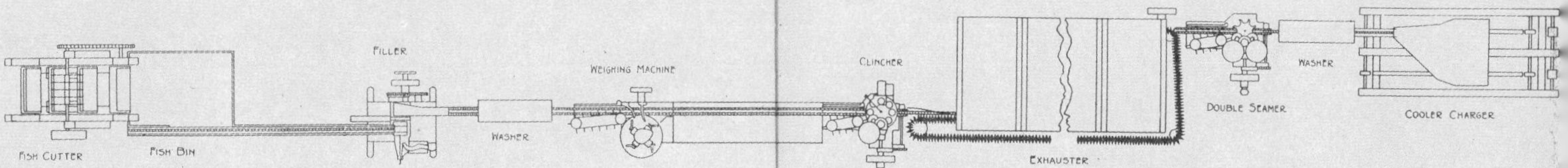

Floor Plan of a Complete Line of Salmon Canning Machinery

ELEVATOR FISH CUTTER

Note the completeness of this line. All the conveyors are self contained, being driven by the machines.

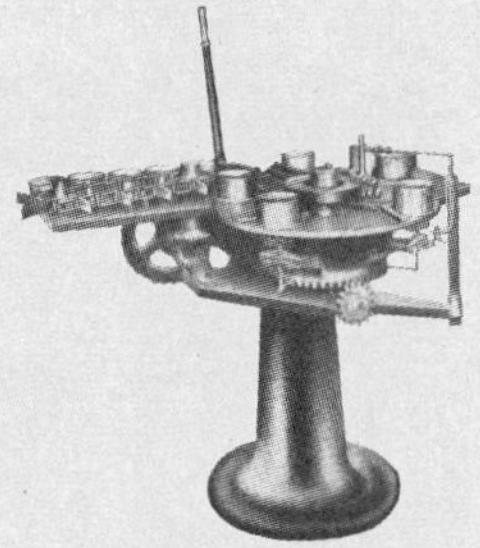

WEIGHING MACHINE
(Patented Aug. 17, 1915; Aug. 3, 1915.)

In this Exhaust Box the cans are carried throughout the entire travel of the box. No dragging to foul and scrape up the dirt. No turntables to cause trouble at the end of the runs. No change required to handle the different sizes of cans.

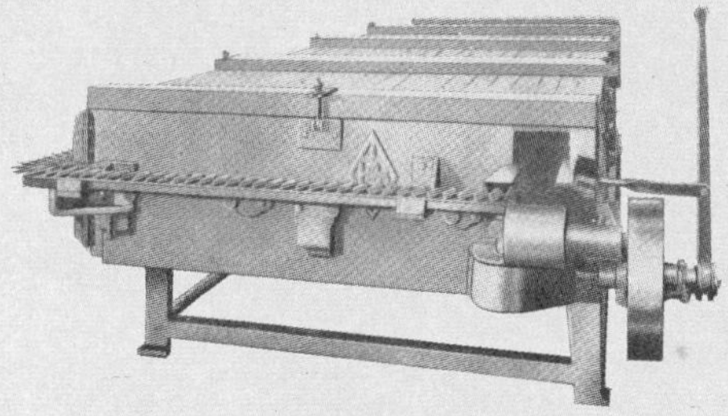

ENDLESS DIAMOND CHAIN EXHAUST BOX
(Patented Sept. 15, 1914.)

All of the Above Machines Are Manufactured by BURPEE & LETSON, Ltd., SO. BELLINGHAM, WASH. LETSON & BURPEE, Ltd., VANCOUVER, B. C.

REPRINTED FROM PACIFIC FISHERMAN YEAR BOOK, 1915, PP. 71-72

RETORTS

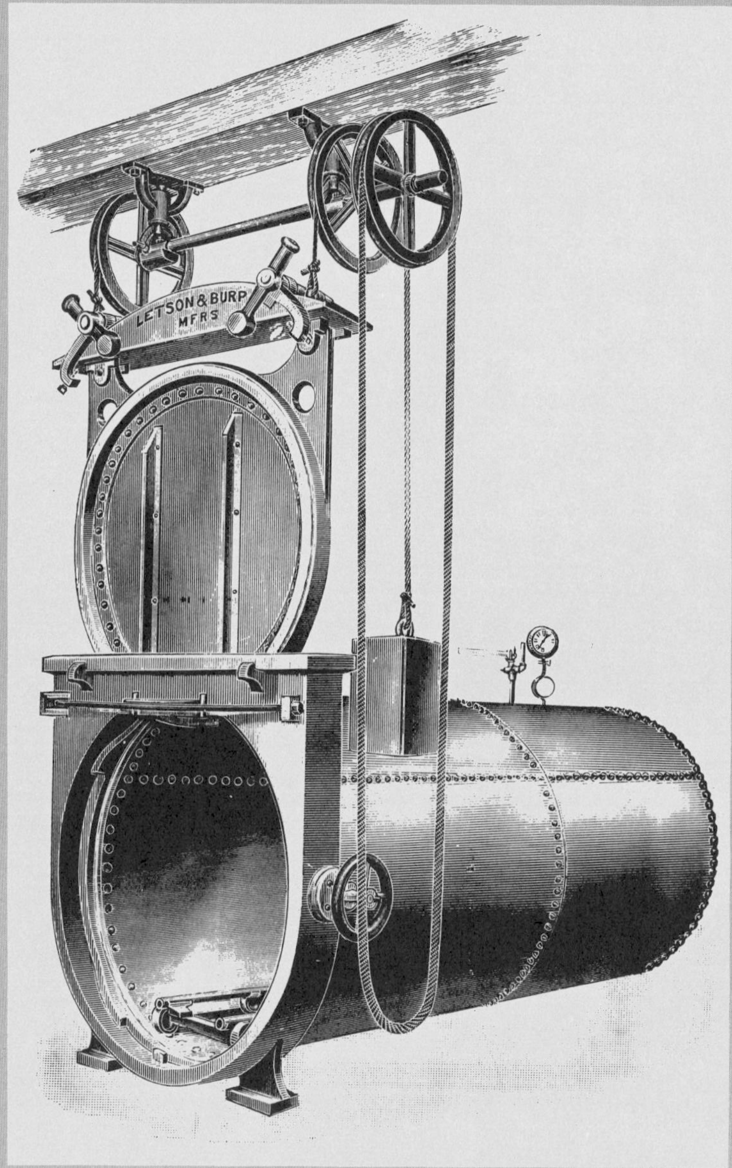

THIS retort is the only retort that is worked out on the correct principles. Instead of forcing the door against the pressure of steam from the outside (as it done by all other retorts on the market) the door withe packing is so arranged as to utilize the steam pressure to make a perfectly steam-tight joint as soon as the steam is turned on.

BURPEE & LETSON, Ltd.,
So. Bellingham, Wash.

LETSON & BURPEE, Ltd.,
Vancouver, B. C.

Agents for the Orient: FRAZAR & CO., 50 Church St., New York City

REPRINTED FROM PACIFIC FISHERMAN YEAR BOOK, 1915, P. 70

In the early 1900's a large battery of retorts was required for cooking. It was the custom to cook in small lots because lines were slower and steam capacity was probably a problem. Each of these retorts had counterweighted doors, and held four cars of cans, as compared with the doubled-ended 10-car retorts used today. Salmon was cooked in the retort under the same principal as the home pressure cooker. Millions of cans of fish were safely cooked in these early retorts over a period of many years without apparent need for present-day government regulations.

COURTESY OF WALTER YONKER

Baranoff Packing Company cannery at Redfish Bay, 1897, on the west coast of Baranoff (Baranof) Island and 58 miles south of Sitka. The first pack was made here in 1891, and capacity was 500 cases per day. This cannery is no longer in existence.

REPRINTED FROM MOSER,
SALMON FISHERIES OF ALASKA, PLATE 46

REPRINTED FROM MOSER, SALMON FISHERIES OF ALASKA, PLATE 49

Above—*Killisnoo, on northeastern side of Kenasnow (Killisnoo) Island, close to the western shore of Admiralty Island, opposite Peril Strait and about 3 miles south of the entrance to Kootznahoo Inlet. The works of the Alaska Oil and Guano Company were located here, for the purposes of extracting oil from herring and converting the refuse to guano. A plant for extracting oil was installed in 1882 and the plant for the manufacture of guano was installed in 1884, both were operated by the Northwest Trading Company until 1888 when the Alaska Oil and Guano Company was formed.*

Right—*Cannery at Pyramid Harbor was built by the Northwest Trading Company in 1883, and located on the western side of Chilkat Inlet, 1½ miles south of Pyramid Island. In 1888 it was sold to D. L. Beck & Sons, San Francisco; in 1889 it burned, but was rebuilt immediately and a pack was made that year. In 1892 the cannery was a member of the Alaska Packing Association, and in 1893 it joined the Alaska Packers' Association.*

REPRINTED FROM MOSER, SALMON FISHERIES OF ALASKA, PLATE 48

COURTESY OF WALTER YONKER

In the early 1900's canneries did not have mechanical elevators or flumes for moving fish, thus salmon were pitched by hand from the floor to the machine.

SANITARY CAN WASHER

For Packers Using Sanitary Cans

U. S. Patent No. 1,178,166, April 4th, 1916

INTERIOR VIEW OF SANITARY CAN WASHER
Floor space required, 10x2 feet; Pulley, 16″ diameter, 2¼″ face
Delivered six cans to one revolution

The Sanitary Can Washer was invented by a practical canneryman, is simple of construction, low in cost and economical in operation. It is connected up to the delivery of double seamers and cans are cleaned before they go into the coolers to be cooked.

The solution of boiling water and soap powder is forced through perforated pipes in the form of spray against the cans while passing through the washer. The cans revolve continuously and every part comes in contact with the spray.

Jensen Can Filling Machine

For One-Pound Tall Cans

This is the original Jensen Filling Machine which gave such excellent service for so many years, with additional improvements which we have perfected in the last three years, including the fork and gate attachment, which does away with all "slam-bang" of the machine and makes it much easier "timed"; the can slide attachment operated by a small cam on the main shaft and the feed barrel spring attachment, which operates the main spring, where it is easily adjusted and gives a direct pull on the feed barrel. Several other features have been added which makes the filler the quietest, smoothest running machine on the market. Fillers are usually furnished with the Gheen patent salter attached. This salter places an equal amount of salt in each can just before it is filled and is generally specified by all canners.

Seattle-Astoria Iron Works

601 Myrtle Street ***(Take South Park Car)*** ***SEATTLE, WASH.***

REPRINTED FROM PACIFIC FISHERMAN, APRIL 1917, P. 48

REPRINTED FROM MOSER, SALMON FISHERIES OF ALASKA, PLATE 51

REPRINTED FROM MOSER, SALMON FISHERIES OF ALASKA, PLATE 52

Above—*Cannery at Eyak Lake, 1897.*
Left—*Cannery at Kussilof (Kasilof), Cook Inlet, 1897.*

REPRINTED FROM MOSER, SALMON FISHERIES OF ALASKA, PLATE 50

Cannery at Orca, about 2½ miles northeast of Cordova. The Pacific Steam Whaling Company moved the cannery to this site in 1895, from the nearby original site, established in 1889.

REPRINTED FROM MOSER, SALMON FISHERIES OF ALASKA, PLATE 53

Above—*General view of Karluk, Kadiak (Kodiak) Island, showing spit, Karluk River, lagoon and village. The 1899 report for the fishing grounds of the Karluk region stated: "The waters surrounding the outlet to Karluk Lagoon are probably the most remarkable in salmon production in Alaska, not only in point of numbers, but in the length of the runs."*

Below—*APA Cannery at Chignik Bay, on southeast shore of Alaska Peninsula.*

REPRINTED FROM MOSER, SALMON FISHERIES OF ALASKA, PLATE 55

Editor's note: *During the first World War, like World War II, the country looked to the salmon industry for food, equipment and experienced seamen, and in both wars the salmon business and its employees made their contributions to the welfare of the nation.*

From: *Pacific Fisherman,* October 1917, p. 11.

A Message From the United States Food Administration to the Fisheries and Fishermen of the Pacific Coast

(By Telegraph)

A full and complete mobilization of the fisheries and speeding up of production is vitally necessary to augment our national food supply and help win the war for democracy. The joint larder of North America and our allies is rapidly lowering: already most of the civil population of Europe are on rations. Our allies have already been compelled to kill over thirty-five million head of their stock animals. Our own exports of meat and dairy products and increased consumption of poultry are reducing these available supplies to the lowest ebb.

Fish is the only remaining flesh product and even in this industry we are now exporting millions of pounds right up to the second line trenches. Partly because of natural causes and partly due to war withdrawals of men and vessels the Atlantic coast catches are far below normal.

Thus because of war conditions the fish industry of the Pacific is now flung into the forefront of national demands. Of all the flesh foods the fish industry alone has the opportunity to greatly increase production, for the crop is inexhaustible and is limited only to the extent of the available vessels, gear and men applied to operations. Every additional million pounds of fresh food fish beyond the normal supply hereafter landed at Pacific Coast ports will release an almost equal amount in meats and dairy produce to add to the rations of our allies.

Let your answer from now on be action—to every last man and every last unit of gear the same spirit of patriotic sacrifice which imbues our armies now fighting for the safety of all civilized firesides. We fully believe that the fishermen of the Pacific will respond as one man to the pressing demands of the hour, and in co-operation with our national food administrators in each state will sound full speed ahead and keep all the spark-plugs sparking. Let the war slogan of the industry, ''Speed up for Uncle Sam'' be nailed to the mast-heads and sound along the Coast from the Gulf of California to Portlock Bank.

UNITED STATES FOOD ADMINISTRATION,
By Kenneth Fowler.

From: *Pacific Fisherman,* October 1917, p. 12.

RESOLUTION PASSED BY THE WASHINGTON CONFERENCE.

"WHEREAS, The United States is engaged in a war in defense of mankind's most treasured possession, the right of all peoples of the world to govern themselves and be free and equal, war in which the conservation of our every resource is vital to its successful conclusion, and

"WHEREAS, The conservation of meats and other non-perishable foods is essential in order that our armies and the armies of our allies be adequately supplied, by the substitution of fish and other perishables to be consumed at home, and

"WHEREAS, Mr. Kenneth Fowler, under the direction of the Food Administration, recognizing the importance of the conservation of fish, has called us, the representatives of the fishing industry, from the various sections of the United States and by doing so has honored us by seeking our cooperation and aid; therefore, be it

"RESOLVED, That we, the conferees, express our appreciation of Mr. Fowler's action in calling us together and giving us this opportunity to co-operate with Mr. Hoover and the Food Administration in the objects set before this conference, and be it further

"RESOLVED, That we take this opportunity and means of expressing our hearty approval of the appointment of Mr. Fowler to be in charge of the fish section of the Food Administration's work because of his practical knowledge and fitness for the work, and be it further

"RESOLVED, That, as patriotic citizens of the United States, we herewith pledge our loyal and undivided support to Mr. Fowler and the Food Administration in the carrying out of their policies."

COURTESY OF PACIFIC FISHERMAN

Bumble Bee Seafoods facility at South Naknek. In the early 1920's, as the salmon business progressed, larger and more modern plants were built and interest from the Columbia River packers increased.

From: *Pacific Fisherman,* October 1917, p. 24.

THE FISHERIES UNDER GOVERNMENT CONTROL.

PRESIDENT WILSON on Oct. 10 issued a proclamation directing the Food Administration, after Nov. 1, to license the manufacture, storage, importation and distribution of some twenty primary foodstuffs. The list includes specifically fresh and frozen fish, and Canned Salmon and sardines; so that, while some minor branches of the industry may not be affected, all important fisheries are included under the rule.

According to a summary of the Food Control Act shown in Bulletin No. 41 of the National Canners' Association, section five provides that, "Whenever the President shall find it essential to license the importation, manufacture, storage, mining or distribution of any necessaries," in order to carry out the purposes of the Act, and shall so announce publicly, no person shall engage in any business specified by the announcement without first obtaining a license. The President is authorized to require accounts to be kept and reports to be made under oath, with full powers of inspection wherever and whenever he deems it necessary; and whenever any storage charge, commission, profit or practice of any licensee is unreasonable, unfair or wasteful, the licensee shall be ordered to discontinue the abuse. The Act also gives the President absolute power to determine what is a just and reasonable charge, etc.

From this it is seen that the President, and through him the Food Administration, has practically unlimited power to dictate regarding prices and business practices when once the license system is in effect. The actual administration of the food control, however, has been placed in the hands of practical men, who have given abundant evidence of their intention to act with caution, and to interfere as little as possible with established trade channels and practices. The machinery for the control of canned foods, including salmon, etc., is already well organized, and its policies fairly established; though control of a somewhat more positive nature may be expected under the licensing system. The organization for handling the fresh and frozen fish department is now being perfected, under the leadership of Kenneth Fowler, a man of high standing and long practical experience in the trade. The vast majority of men in both departments of the industry have shown full realization of the need for control, and are readjusting their trade organization and methods for loyal support of Food Administration policies.

The reason for placing the industries under license is not to compel acquiescence in anything detrimental to the interests of those legitimately engaged, but, as in the case of conscription of men, to protect the patriot against the slacker: to enable those who wish to strengthen the nation's arm in war to do so with the assurance that their patriotic action will not give unfair advantage to the few who might hold back. This is now generally understood, and the application of the licensing system is done with the hearty approval of the great bulk of the trade.

NO SUSPENSION OF SALMON CONSERVATION LAWS.

Some uncertainty as to the intentions of the Food Administration in requesting the suspension of local restrictions on fishing operations is cleared up by a wire from Dr. Hugh M. Smith, head of the U. S. Bureau of Fisheries, received by Pacific Fisherman Oct. 12, stating that the Bureau of Fisheries had concurred in the recommendation that local laws be relaxed or suspended as a war measure to secure increased output of salt water fishes, but that this action by the Bureau is not intended to affect Pacific salmon fishing or to permit the entrance of alien fishermen. This supports the impression that in making this general recommendation the Food Administration had in mind certain Eastern and Middle Western districts rather than the Pacific Coast, and that no very radical change in fishery regulations is desired in this section.

Waterfall cannery on west coast of Prince of Wales Island was owned by Nakat Packing Company. It was a trap and seine fishing operation. With the demise of the traps this immense and very modern facility was sold at a bargain price ($50,000) as a resort.

COURTESY OF TED FRIELE

From: *Pacific Fisherman*, March 1917, p. 17.

IMPORTANCE OF PROPER FUEL TANKS AT CANNERIES.

A matter which has been coming more and more to the attention of salmon packers during the last year or two is the use of substantial and properly placed tanks as containers for fuel oil, distillate, etc., at the canneries, and dispensing with the drums which have heretofore been too generally used. As was emphasized in an article in Pacific Fisherman for February, the use of suitable tanks, with various precautionary measures, is strongly urged from the point of view of fire hazard, as serious losses have occurred through the use of leaky, defective or improperly placed tanks or drums.

Insurance men who have had wide observation of cannery fires dwell on the point that containers, especially of distillate or gasoline, should be so placed that there is no danger of a flow of the inflammable liquid to start a fire or make it more intense. It is also important that tanks be so placed that there is no danger of their being carried away by flood or other disturbance; a case being mentioned in which a tank located 500 feet from a cannery was lifted by a flood, permitting the gasoline to leak and float down to the cannery, which was destroyed through the ignition of the floating gas. The placing of tanks underground is advised wherever possible; but in the case of salmon canneries, using frequently tanks of 15,000 gallons capacity, this is obviously impracticable. The tanks, however, should in all cases be placed on solid ground at least 50 feet from the cannery, and should be surrounded by a dike or ditch to catch the oil in case of leakage or bursting of the tank.

In the use of drums, the danger from leakage or spilling of distillate or oil is multiplied many times, while the likelihood of their being carelessly left at points where the risk of ignition is great makes them a constant source of danger. Aside from the fire risk, however, canners are finding the use of drums wasteful, and many are already figuring on replacing them with tanks. On this basis it is estimated that the tanks will pay for themselves in about two seasons, as they will effect a saving of nearly 7 per cent as compared with the cost of handling .the oil or distillate in drums. In the first place, the use of drums usually involves a waste of from 3 to 5 gallons in each drum; and the labor saved by pumping into a tank rather than handling drums over the wharf is an item of considerable importance. In addition to this is the 1c per gallon charged on returned packages, which amounts to considerable money in a season.

Many of the canners have been anxious to put in tanks this sasoen; but unfortunately the steel situation and the rush of work in the shops makes this almost impossible, and most of them will delay action until next year, when the movement toward the substitution of tanks for drums will probably be quite general.

From: *Pacific Fisherman*, October 1917, p. 15.

HEARINGS ON CANNERY LABOR CONDITIONS

Washington State Labor Commissioner C. H. Younger and L. Woods, U. S. district employment agent at Seattle, spent Sept. 19 and 20, taking the testimony of some forty laborers returning from the Bering Sea canneries, in regard to complaints made by the men concerning their treatment and the methods of payment on reaching port. The complaints did not directly involve the packing companies, but were directed against the Chinese labor contractors and their foremen.

A large amount of testimony was taken, which will, it is said, be called to the attention of the Federal authorities, and will also be used by the State labor commissioner as a basis for proposed legislation, modeled in a general way after that now in effect in California, governing the treatment of labor sent out from the state. Neither Mr. Woods nor Mr. Younger has assumed any jurisdiction in the case, but a number of the laborers have engaged an attorney and threaten civil action.

Some of the cannerymen express willingness to co-operate to secure legislation that might correct some of the troubles inherent in the present system; but little hope is felt of making any great improvement in conditions unless a larger supply of reliable labor can be made available.

From: *Pacific Fisherman*, April 1917, p. 16.

BRISTOL BAY PACKING CO. PLANS NEW CANNERY.

The Bristol Bay Packing Co. will have a new cannery in addition to the old one on the Koggiung river this season and their capacity will be more than doubled. This plant is to be equipped with the latest sanitary can machinery and other improvements. The capacity of the old cannery and the new addition will be about 150,000 cases. The crew to be sent north this season numbers about two hundred fishermen, about one hundred mechanics and 285 Orientals and Filipinos, thus making the force close to six hundred. The company has five carloads of tinplate on the way from the East, but as this is only one-fourth of the full requirement, some anxiety is felt about the remainder being delivered on time. The fleet of the company consists of: The steamer "Costa Rica," the bark "McLaurin," the bark "Narwhal," and the barkentine "City of Sydney." The last is one of the largest vessels in the industry, having a cargo capacity of 2,679 net tons. This vessel is now on her way back from a voyage across the Pacific. The sailing vessels will start north April 15, and the steamer on May 1.

ICY STRAITS PACKING CO. ORGANIZED.

The new project mentioned in December as being planned by J. C. McBride, Herman T. Tripp and others is now taking shape, the company having been incorporated at Juneau with a capital of $100,000, under the name of the Icy Straits Packing company. Mr. McBride is president and Mr. Tripp vice-president, with H. L. Faulkner as secretary-treasurer.

ADDITION TO HERENDEEN BAY PLANT.

The Bering Sea Packing Company, controlled by the same interests as the Everett Packing company, on March 30 dispatched the power schooner Ruby from Seattle with lumber and supplies for the addition to the plant on Herendeen Bay, which was built last year. This company has also chartered the power schooner Bender Bros., which recently returned from a trip to Morzhovoi Bay for the Sockeye Salmon company. The Bender Bros., after undergoing repairs, left for the cannery early in April. Both vessels are expected back on the Sound before the middle of May.

From: *Pacific Fisherman*, January 1917, p. 17.

NEW CANNERY FOR ALASKA.

It is reported that D. N. McMillan, of Bellingham, Wash., and his two sons Joseph and George, will establish a salmon cannery on False Pass, Alaska peninsula, in time to operate during the season of 1917. Joseph McMillan is in that region at present looking over the ground. The McMillans were interested in the Bering Sea Packing Co., which built and operated a cannery on Herendeen bay in 1916, but it is reported that they have disposed of their interest in that company.

PACIFIC AMERICAN FISHERIES GIVES WAGE INCREASE.

The Pacific American Fisheries, which operates canneries on Puget Sound and in Alaska, on Dec. 23 announced a voluntary wage increase of 10 per cent for 1917 to all employes receiving less than $1,500 a year and a Christmas gift of $10 to each employe of the company. Several thousand workers will be affected by this generous act of the company.

COURTESY OF NATIONAL FISHERMAN

LARRY FREEBURN

Top—*A trap watchman's shack for a floating trap. It contained two bunks, a shiplap table and a stove. The watchman kept kelp and debris out of the trap gear and was a deterrent to unwelcome callers.* Above—*A rigging scow.*

Right—*Pile trap at Kelly Rock, Squaw Harbor, Alaska.* Below—*Preparing to take wire out to trap site.*

COURTESY OF MRS. ANDREW GRONHOLDT

LARRY FREEBURN

Right—*The floating trap, devised in 1910 by Bob Heckman, enabled packers to put traps in areas where water was too deep for pile traps. This trap was the mainstay of the industry, permitting controlled production and high quality. Floating traps were outlawed in 1959.* Below—*The last brail. An apron-type brail allowed the scow to be loaded quickly.*

H. C. SCUDDER

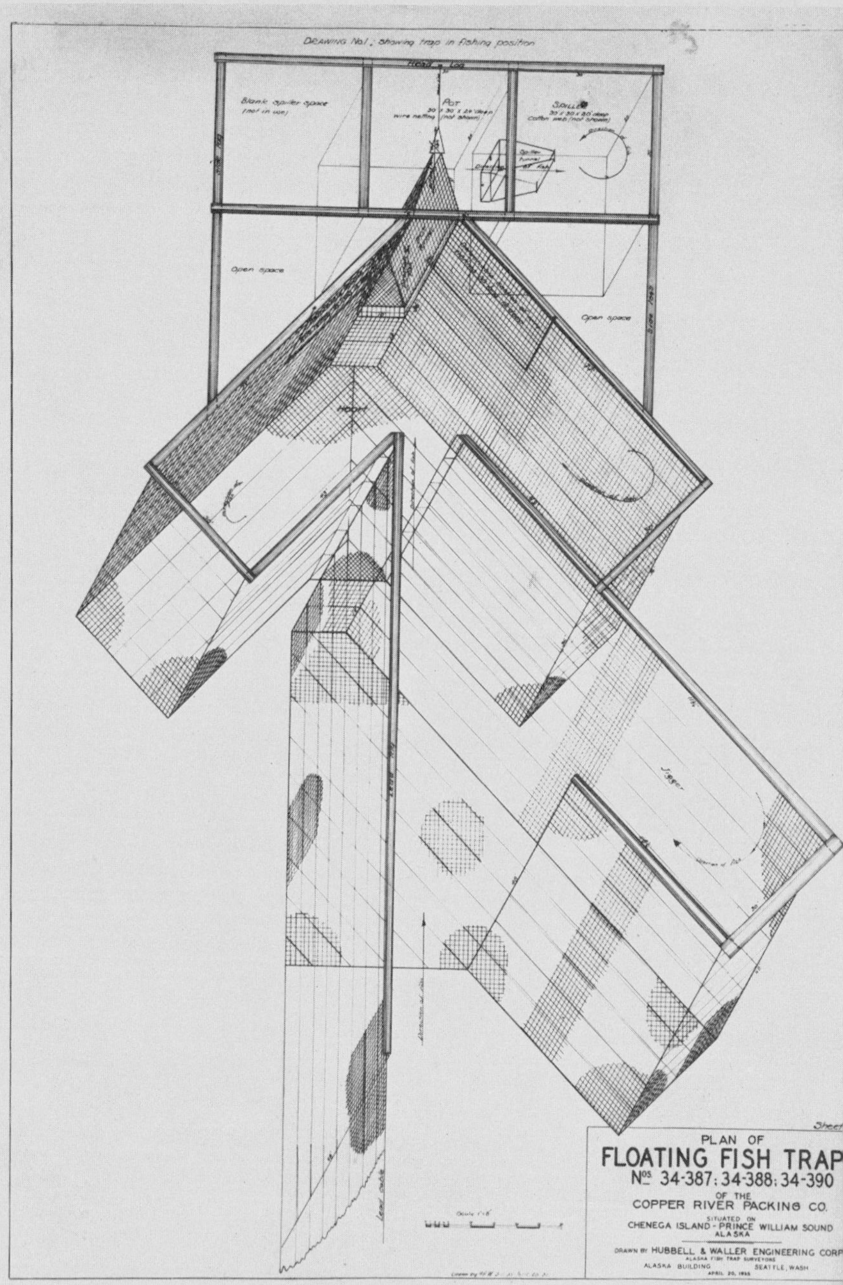

COURTESY OF LARRY FREEBURN

Members and Guests of the Association of Pacific Fisheries, Assembled at the Fifteenth Annual Convention, Vancouver, B. C., October 20, 1928

Front Row, Left to Right—S. H. Cooke, author of "Bunk House Steps" stories in Pacific Fisherman; Leo White, Oceanic Sales Co.; Wm. C. Graham, Pres. Northern Electric Co., Seattle; James Cannon, Nakat Packing Corp.; Chas. D. Garfield, Alaska Dept., Seattle Chamber of Commerce; John N. Cobb, Dean of the College of Fisheries, University of Washington; August Buschmann, Pres. Deep Sea Salmon Co., Third Vice-President Association of Pacific Fisheries; J. F. Freeburn, Pyramid Packing Co.; J. P. McElhany, Sec.-Treas. Peril Straits Packing Co.; E. I. Bentley, Pres. California Packing Corp., First Vice-President Association of Pacific Fisheries; Victor H. Elfendahl, Vice-Pres. Sunny Point Packing Co., President Association of Pacific Fisheries; H. W. Phelps, Pres. American Can Co.; Dr. Ernest D. Clark, Director Northwest Branch, National Canners Assn., Secretary Association of Pacific Fisheries; Fred P. Kendall, Northwest Manager American Can Co.; R. A. Welsh, Jr., Bellingham Canning Co.; J. A. Connolly, G. P. Halferty & Co.; R. A. Welsh, Pres. Bellingham Canning Co.; Paul E. Derby, Tropical Paint & Oil Co., Seattle.

Second Row, Left to Right—Joe Most, Sec.-Treas. Copper River Packing Co.; E. B. Hanley, Pres. Copper River Packing Co., Fourth Vice-President Association of Pacific Fisheries; W. W. Scruby, Vice-Pres. Dexter Horton National Bank; R. R. Farish, Seattle Manager New England Fish Co.; Bert Cochran, Seattle Manager H. K. McCann Co.; Egil Buschmann, General Supt. Nakat Packing Corp.; Leif C. Buschmann, Deep Sea Salmon Co.; A. L. Hager, Vice-Pres. and General Manager, New England Fish Co., Pres. Canadian Fishing Co., Ltd.; William Timson, Pres. Alaska Packers Association; C. A. Burckhardt, Pres. Alaska Consolidated Canneries; G. P. Halferty, Pres. Pioneer Packing Co. and G. P. Halferty & Co.; Thomas E. Collins, Street Car Advertising Co.; Wm. Prosser, Strong & Prosser, Seattle; Wm. Coleman, Blanchard, Coleman & Nichols; Robert R. Payne, Production Manager Canadian Fishing Co., Ltd.; Charles R. Pollock, Supervisor of Fisheries, State of Washington.

Third Row, Left to Right—Crescent P. Hale, Pres. Alaska Salmon Co., and Bristol Bay Packing Co.; John Dorman, Mgr. Bones Bay Cannery, Canadian Fishing Co., Ltd.; F. J. Comeau, F. J. Comeau & Co.; N. D. Jarvis, College of Fisheries, University of Washington; O. E. Shostrom, Chemist, Northwest Branch, National Canners Assn.; Wm. T. Hale, North Pacific Trading & Packing Co.; L. J. Hull, Pres. Alitak Packing Co. and Oceanic Sales Co.; Dr. R. W. Clough, Chemist, Northwest Branch, National Canners Assn.; Geo. C. Teal, Secretary Pacific Canned Salmon Brokers Assn.; Nick Bez, Vice-President Peril Straits Packing Co.; J. F. McGovern, McGovern & McGovern, Pres. Pacific Canned Salmon Brokers Assn.; B. T. Weyand, Seattle Mgr. R. E. Cotter Co.; Frank E. Derby, Tropical Paint & Oil Co., Seattle; Carl A. Sutter, General Mgr. [illegible] Island Packing Co.; [illegible] to the President, Union Pacific System; W. H. Olin, Asst. Gen. Freight and Passenger Agt. O. W. R. & N. Ry.; Stedman H. Gray, Editor Pacific Fisherman; [illegible] Correspondent Pacific Fisherman.

Rear Row, Left to Right—Frank E. Gorrell, Secretary National Canners Assn., Washington, D. C.; D. W. Branch, General Mgr. Salmon Dept., Libby, McNeill & Libby; [illegible] Mgr. Wales Island Cannery, Canadian Fishing Co., Ltd.; Alex. Rutherford, Mgr. [illegible] Bay Cannery, Canadian Fishing Co., Ltd.; W. S. Dukes, District Sales Mgr. American Can Co., Portland, Ore.; H. A. Taylor, Plymouth Cordage Co.; F. B. Andrews, Pacific Marine Supply Co.; M. M. Houck, Nakat Packing Corp.; John Richardson, [illegible]; A. B. Whittle, Sec. Columbia River Packers Assn.; G. L. Wikstrom, Seattle Mgr. [illegible]; Hon. Wm. Steele, Alaska Territorial Senate; Wm. C. [illegible], Pres. [illegible] Packing Co.; M. J. Sullivan, San Francisco, Vice-Pres. American Can Co.; T. R. [illegible], Mgr. [illegible] & Manitou Canneries, Canadian Fishing Co., Ltd.

Above—*Association of Pacific Fisheries convention, October 20, 1928.* Right—*Prominent members of the salmon industry at Fishery Advisory Council, 1936, in Washington, D.C. From left, third man is Frank Bell, Commissioner of Fisheries; fourth, Charles Jackson; fifth, Bert McGovern; eighth, Nick Bez; and tenth, C. J. Sebastian.*

COURTESY OF JERRY OAKSMITH

PHOTOS COURTESY OF JERRY OAKSMITH

Clockwise from above—*The can body goes into the reform machine to be shaped into a cylinder. After can is flanged the bottom is put in. Downstairs the "Iron Chink" beheads the fish, clips the fins off and removes the viscera.*

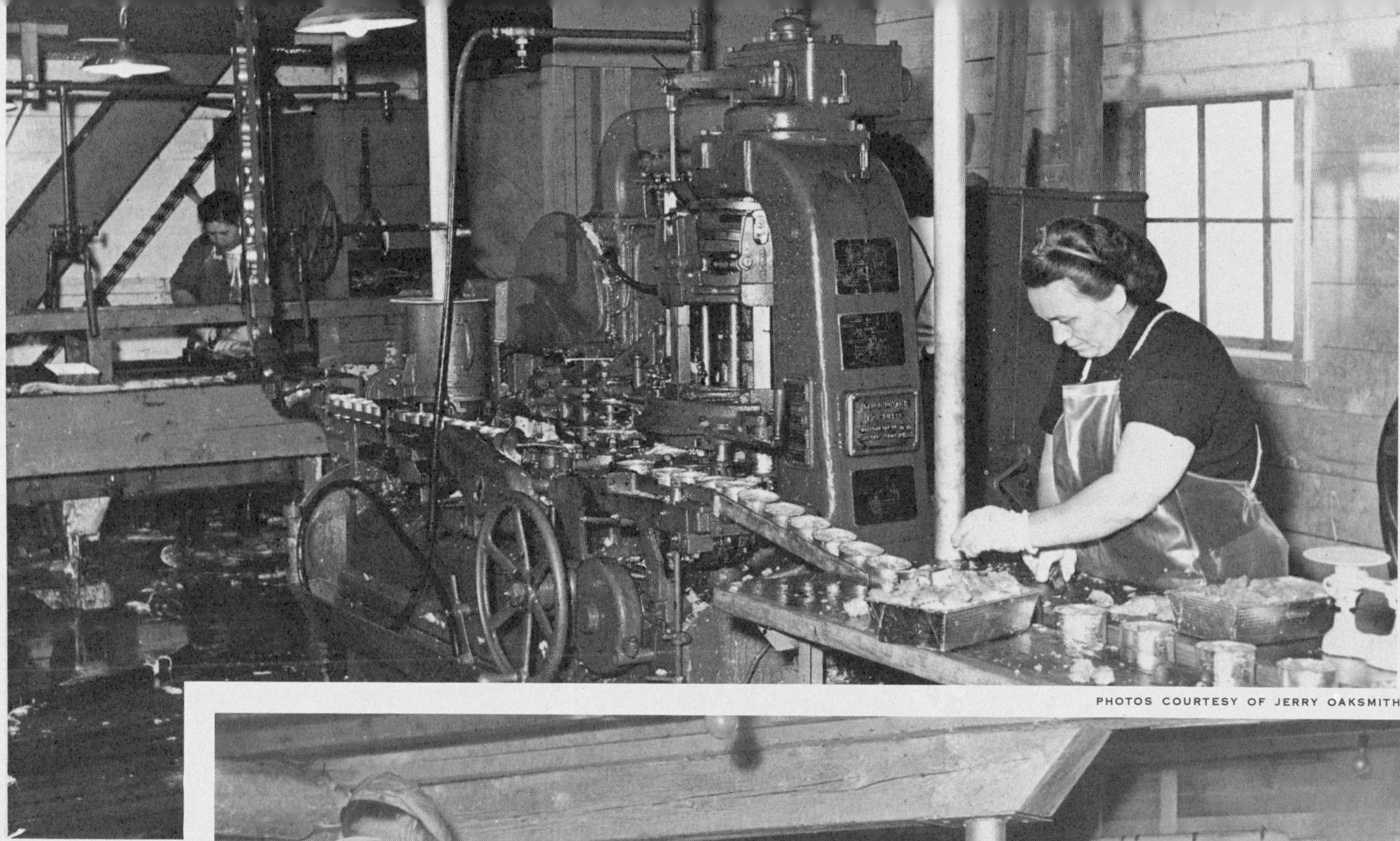

Left—*The cans are filled by machine then checked for appearance and weight.* Below—*Ready to cook in the retort. All photos on these pages are of a small cannery operation at Seldovia in the middle fifties.*

PHOTOS COURTESY OF JERRY OAKSMITH

SEWARD FISHERIES

Above—*Seward Fisheries' modern plant at Seward is diversified for all seafood products. Few plants have been built in recent years, and as a result of the decline of the salmon industry other products are being handled. Access to transportation facilities and availability of some local help have been determining factors in selecting sites for plants built in the past few years, and none have been constructed in isolated areas.*

Wrapped in tin and ready to cook.

LARRY FREEBURN

COURTESY OF BOB SYRE

Left—*Excursion Inlet Packing Company, a modern canning and cold storage plant located on inlet of same name, 38 miles west of Juneau.* Below—*King Cove cannery, owned by Peter Pan Seafoods, is about 18 miles southeast of village of Cold Bay on southwest coast of Alaska Peninsula. In spite of constant closing of salmon canneries, large modern plants, such as this, handling diversified products (salmon, halibut, crab), and surrounded by local help likely will remain in existence.*

COURTESY OF JON JOHNSON

SALMON—OUR GREATEST FOOD FROM THE SEA

1. The salmon canning industry in Alaska doubled in volume with the advent of pile-driven traps such as the one shown above. They are licensed, taxed, and closely regulated.

2. Floating traps are anchored. They came into general use about 1917 in localities where the bottom did not permit pile driving. All traps are closed during week-end periods.

3. The brailing of salmon into scows makes for economy and ease of operation, both in taking fish from the traps and unloading at the cannery.

4. The advent of the Iron Chink for beheading and "cleaning" the salmon marked a great advance in the salmon canning industry.

5. After it leaves the Iron Chink every salmon is individually washed, inspected and doused continuously with fresh, running water.

6. To save space cans are shipped north in a collapsed condition and are re-formed at the cannery by machines such as above.

7. The cans are salted and filled with one pound of salmon by speedy machines like the Filler above.

8. An automatic Weighing Machine discards cans which are even as little as one-eighth of an ounce under the full pound weight. Such cans are later filled by inspectors wearing gloves who examine all cans as they pass their tables.

- **For thousands of years, hundreds of millions of salmon ascended the streams and rivers of Alaska each autumn to spawn. They died and were of no use to man. Then, in 1878, at Klawock, the Canned Salmon Industry made a modest start.**
- **With the steady increase in the activities of Salmon Canning, Alaska grew with the industry. Capital which Alaskans did not have, poured in, and cities grew. Alaskans appreciate these activities which have grown from their great resource.**
- **Consumers in the States and all over the world know Alaska salmon for a nutritious, quality food, easily prepared in many different ways, or eaten just as it comes from the can. Salmon canning is Alaska's greatest industry, and tax revenues derived from it are the chief support of Territorial Institutions and schools.**

9. The Clincher clinches a top to the can. The cans then go to the Vacuum machine which draws out the air and seals the can. The cans of salmon are now ready for cooking.

10. The cans are cooked in Retorts, like the above under steam pressure.

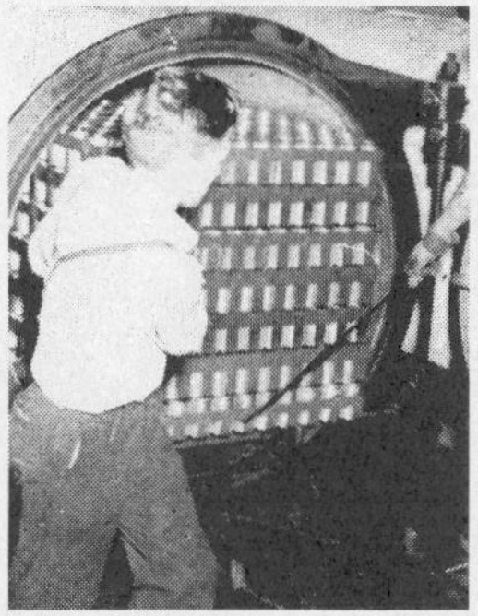

11. After being cooked for approximately one hour and twenty minutes, the cans are cooled and washed before being labeled.

12. A conveyor and turntable sends the cans to the labeling machine.

14. During rush times, cans are sometimes stacked by the thousands before being labeled.

13. The cans are packed into a carton by machine and the cartons are automatically sealed.

15. Southbound ships call at the cannery and the ships are loaded with the finished product for shipment to "the States."

OLLOWING SALMON PACKING CONCERNS:

Hydaburg Cooperative Association
Independent Salmon Canneries, Inc.
Alaska Seldovia Packers, Inc.
New England Fish Co.
Northern Commercial Co.
San Juan Fishing & Packing Co.
Whiz Fish Products Co., Inc.

REPRINTED FROM THE ALASKA SPORTSMAN® SEPTEMBER 1947, PP. 34-35

Loading seine after a haul, circa 1912.

JOHN N. COBB COLLECTION, UNIVERSITY OF WASHINGTON LIBRARY

Boats and Gear

PHOTOS FROM JOHN N. COBB COLLECTION, UNIVERSITY OF WASHINGTON LIBRARY

Left—*Four-oar seine boats, about 1911 and prior to the use of gas engines. Photo was taken looking north from Fidalgo Island Packing Company cannery, Ketchikan.* Above—*Ketchikan Creek, Alaska, was leased for seining. Photo was taken approximately 1911.*

PHOTOS FROM S. P. NIELSEN COLLECTION, SAN FRANCISCO MARITIME MUSEUM

Above—*Early day seining was a matter of a skiff, four oars, a net and plenty of hands.* Right above—*Beach seining without power—a fair haul.* Right—*"The last haul"—seining by hand.*

ALASKA PACKERS' ASSOCIATION COLLECTION, SAN FRANCISCO MARITIME MUSEUM

Fishermen lifting an Italian floating trap off live bit (part of a working winch) in Chignik Bay, Alaska, about 1911-12. There were 42 Italians employed on 1 trap, working in 2 shifts. Lead of trap was 3 miles long.

A deckload—1944. The seiners tried to make the author eat these fish during the big run in Lisianski Inlet in 1944. No one was put on limit.

LARRY FREEBURN

JOHN N. COBB COLLECTION, UNIVERSITY OF WASHINGTON LIBRARY

COURTESY OF JOHN GILBERT

Above—*A small launch tends a four-oar seiner, circa 1911.* Left—*A nice set on a small school. This salmon purse seiner operated in Southeastern Alaska.*

Fleet of salmon seiners in early 1920's on a Sunday. Vessels include Comet, Sea King, Rondout, Monarch, Veribus, Juno *and* Minneapolis.

JOHN N. COBB COLLECTION, UNIVERSITY OF WASHINGTON LIBRARY

The purse seiner Urania *was to be rebuilt many times.*

COURTESY OF FRANK BERRY

REPRINTED FROM PACIFIC FISHERMAN, MAY 1917, PP. 30-31

Wisconsin CONSISTENT Motors

All of the qualities you desire in a marine motor you will find embodied in the WISCONSIN in the highest degree. Abundant power, yet light and compact. Accurately balanced—smooth running. Noted for consistent performance. Write for specifications of different types.

WISCONSIN MOTOR MFG. CO. Sta. A Dept. 305 **Milwaukee, Wis.**

Pacific Coast Distributors:
Earl P. Cooper Co., 1428 Bush St., San Francisco, Cal., and 1310 So. Angeles St., Los Angeles, Cal.
W. F. McKenney, 68 First St., Portland, Ore.
Chandler, Dunlap Co., Seattle, Wash.

Type Q. M. 4-cylinder, 4-cycle. Bore 3 ¼″. Stroke 5″.

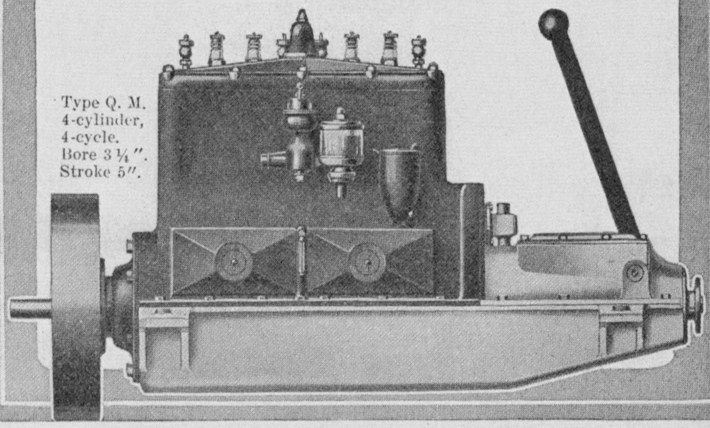

REPRINTED FROM PACIFIC FISHERMAN, SEPTEMBER 1917, P. 43

The ENGINES AND *The* SERVICE

These well-known Standard agencies now handle both

STANDARD DISTILLATE ENGINES

and

SOUTHWARK-HARRIS VALVELESS OIL ENGINES
(Diesel Principle)

Assuring patrons REAL SERVICE in all the principal ports of the Pacific.

STANDARD owners, to avail themselves of *Factory Service*, should "Buy Standard Parts from Standard Agents."

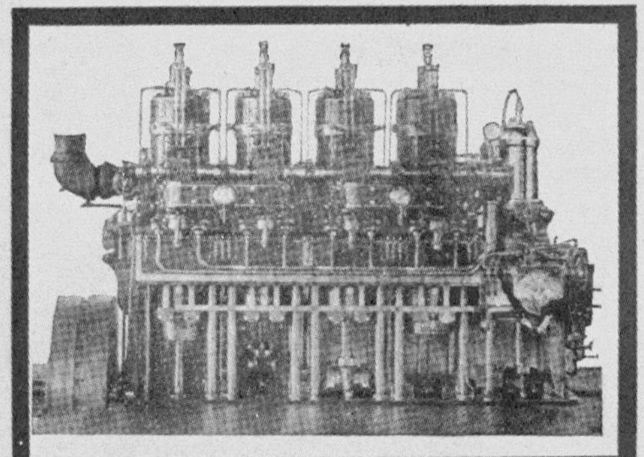

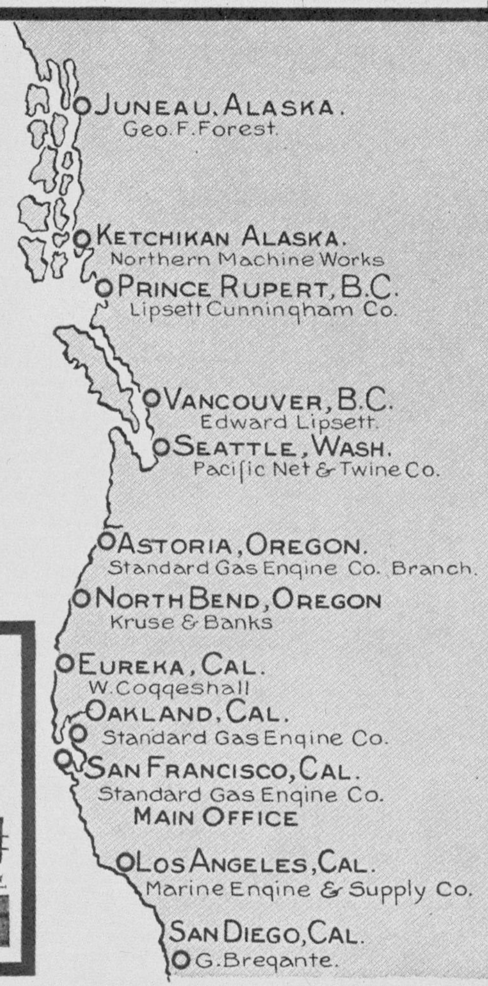

STANDARD GAS ENGINE CO.
SAN FRANCISCO.

REPRINTED FROM PACIFIC FISHERMAN YEAR BOOK, 1915, P. 60

The UNION Is Used by

The Largest Canners—
The Largest Steamship Companies—
The Largest Railroads—
The Largest Towboat Companies—
The Largest Oil Refiners—
The United States and Foreign Governments—
And by Thousands of Individual Owners—
Because They All Know That—

THE UNION IS CERTAINLY
THE BEST BUILT ENGINE

80-600 H. P. OPEN CROSSHEAD TYPE

UNION GAS ENGINE COMPANY Oakland, California

REPRINTED FROM PACIFIC FISHERMAN YEAR BOOK, 1916, P. 108

Tanning linen nets before the season, during the pre-nylon era.

JOHN N. COBB COLLECTION, UNIVERSITY OF WASHINGTON LIBRARY

CAPTAIN PETER LARSEN COLLECTION, SAN FRANCISCO MARITIME MUSEUM

Above—*Stake net at Bristol Bay, circa 1914. This fishing method is no longer used.*

Below—*Dipping salmon from the Copper River, approximately 1911.*

JOHN N. COBB COLLECTION, UNIVERSITY OF WASHINGTON LIBRARY

KATHERINE MARTIN COLLECTION, SAN FRANCISCO MARITIME MUSEUM

Left—*Tired, and no fish—gill-netters, about 1914.*

ALUMINUM GILL NET FLOATS

WHY THEY ARE THE BEST:

1. They are the cheapest because they last the longest. Floats that have been in constant use for the last 15 years are practically as good as new.
2. They never get water soaked and therefore save the fisherman a great amount of labor and can be used continually.
3. They are seamless, guaranteed not to leak and will withstand a pressure of at least 80 fathoms of water.
4. They attract fish because of their bright appearance in the water.
5. They are made of pure aluminum and will not rust or corrode.

These floats are being used extensively by nearly all the fishermen on the Great Lakes and have been proven a big success. They are made in two sizes: 1⅝"x4⅝" and 1¾"x5¾".

Samples and prices sent on request.

ALUMINUM GOODS MANUFACTURING CO.
Manitowoc, Wisconsin.

REPRINTED FROM PACIFIC FISHERMAN YEAR BOOK, 1914, P. 78

COURTESY OF NATIONAL FISHERMAN

"You caught me!" The salt fish for home was important although illegal when taken from under the dock during closed season.

CARL JOHNSON COLLECTION, SAN FRANCISCO MARITIME MUSEUM

Above—*"Whitehead Pete's" (Nelson) gill-net boat in Kvichak River, 1914.*

CARL JOHNSON COLLECTION, SAN FRANCISCO MARITIME MUSEUM

Above—*Southeast gale strikes the Alaska Packers' station at Pilot Point, Ugashik, Alaska.* Left—*With fish in the net, all aboard, and net yet to be picked. Gill-net fishermen in Bristol Bay.*

SAN FRANCISCO MARITIME MUSEUM

Opposite—*Heading for Bristol Bay grounds, belly full of grub ashore and ready to fish.*

J. JOHANNESEN COLLECTION, SAN FRANCISCO MARITIME MUSEUM

73 NC
24
24 NC
NC 5
5
65
65 NC
74
12 NC
NC 23

COURTESY OF HAROLD BRINDLE

Drying out after a storm, 1927. Sails must be dried to prevent rot. The fishermen likely are ashore, resting and waiting for the next fishing period.

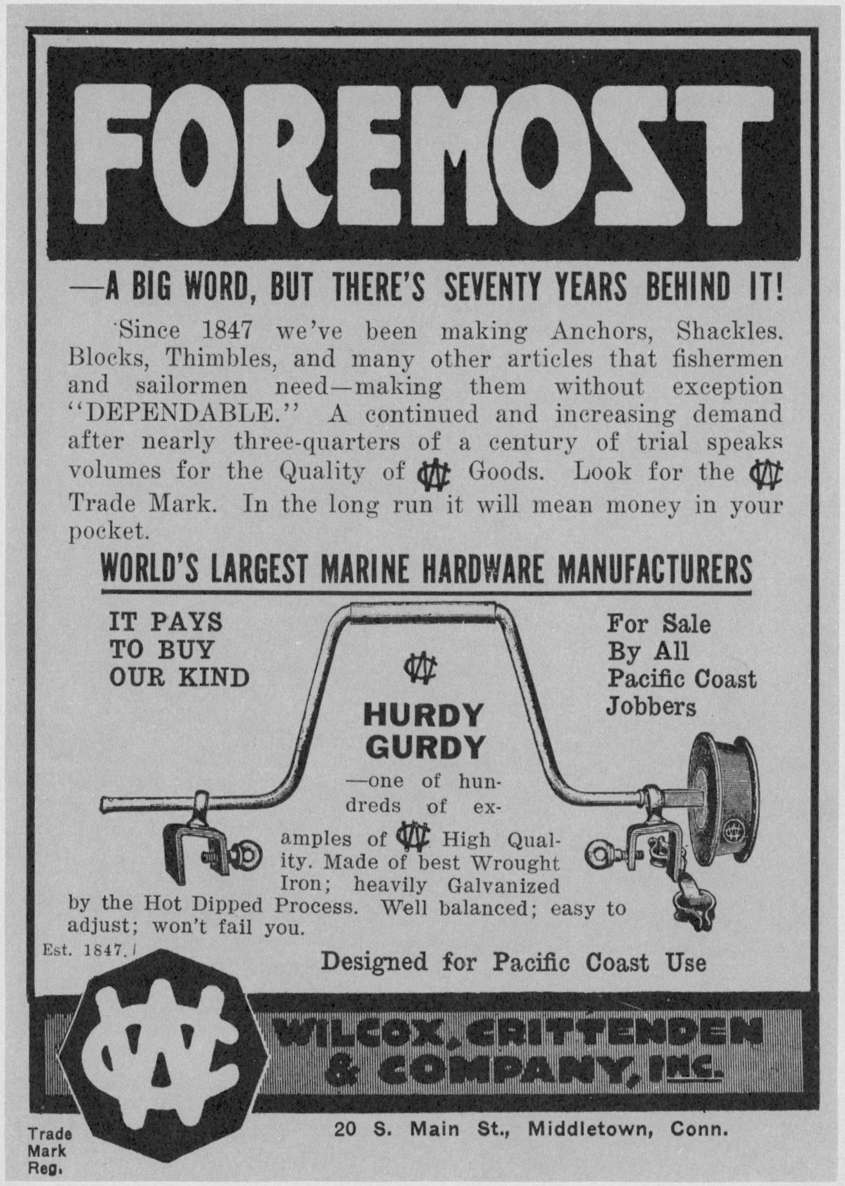

REPRINTED FROM PACIFIC FISHERMAN, SEPTEMBER 1917, P. 51

COURTESY OF HAROLD BRINDLE

Sunday on the beach, 1927. Time for relaxation, reflection on gripes and hopes, and preparation for the next big set.

TARRED LINES OF QUALITY

Burnham's Lines

L. A. Burnham & Co.

TRADE MARK

GLOUCESTER, MASS.

"ASK THE CAPTAIN WHO HAS USED THEM"

BARBOUR'S

Double BARBOUR FLAX SPINNING COMPANY TRADE MARK FLAX Knot

SALMON GILL NETTING

Barbour's & Finlayson's Net Threads

ADVERTISEMENTS REPRINTED FROM PACIFIC FISHERMAN, APRIL 1917, P. 7

From: *Pacific Fisherman,* February 1917, p. 15.

MEDICAL AID FOR FISHERMEN.

The steamer Unalga of the Coast Guard Service arrived at Juneau Jan. 20 and is now on a cruise through the waters of Southeastern Alaska and the Gulf of Alaska for the purpose of rendering medical aid to the crews of American fishing boats.

The Unalga will travel from Sitka to Cape St. Elias and Middleton Island and over Portlock fishing banks. Medical aid may be obtained aboard the vessel in Yakutat when the ship is not cruising over the above routes.

R. G. Dodge, captain United States Coast Survey, is in command of the Unalga.

HARRY HOYER COLLECTION, SAN FRANCISCO MARITIME MUSEUM

1908

COURTESY OF HAROLD BRINDLE

Left—*Scows were built by shipwrights on the beach—1908.* Above—*Alaska Packers' Association cannery at North Naknek and gill-net boats in summer of 1939. Cannery Superintendent Heinbokel posted the wooden star on the side of the building to call in the towboats because there was no radiophone at this time.*

AXEL WIDERSTROM COLLECTION, SAN FRANCISCO MARITIME MUSEUM

COURTESY OF WILHO ALENIUS

COURTESY OF HAROLD BRINDLE

Above clockwise—*Alaska Packers' Association steam tugs* Queen *and* Quinnat *at anchor at Naknek River, 1911. Herman and coffeepot on a drift—"Home sweet home." Mending nets at Naknek Packing Company plant, 1924.*

46

Sunset and no wind—Naknek River, 1946-47.

COURTESY OF JOHN GILBERT

REPRINTED FROM PACIFIC FISHERMAN YEAR BOOK, 1916, P. 48

From: *Pacific Fisherman*, September 1917, p. 25.

BRISTOL BAY SHIPS RETURNING.

The first of the Bristol Bay fleet to reach port after the season's operations was the Alaska Packers' Association's ship Indiana, which reached San Francisco Sept. 4, after a quick and uneventful voyage, with approximately 42,000 cases of salmon. According to statements made by her officers, only adverse weather conditions prevented a pack in Bering Sea far beyond all former records.

The Seattle fleet was led by the ship Benjamin F. Packard, of the Northwestern Fisheries company, which came in Sept. 6, after a voyage of 18 days from the Nushagak. This was the quickest trip she had ever made, and no small credit is given to her master, Capt. H. Bune, for bringing her in safely in such good time. She brought a full cargo from the cannery, besides 225 hands. The Northwestern Fisheries company received word of the departure of the ship St. Paul from Kenai for Seattle Sept. 1, and she is expected at any time. The company's bark Guy C. Goss is due to leave Chignik Sept. 15.

The ship Abner Coburn, of Libby, McNeill & Libby, reached Seattle Sept. 9, bringing 51,000 cases of salmon from the Nushagak. The Coburn also brought 150 cannery hands.

BARK ST. KATHERINE SAVED.

A message was received late in August that the Red Salmon Canning company's bark St. Katherine, which went ashore last May, had been successfully floated and towed to Ugashik for temporary repairs, preparatory to being towed to San Francisco for a thorough overhauling. It was at first reported that she was a total loss, but she suffered comparatively little damage.

A lively breeze and in comes the net.

CLARA W. JAMES PHOTO COURTESY OF JOHN GILBERT

From: *Pacific Fisherman*, September 1917, p. 25.

STORMY SEASON AT BRISTOL BAY.

Old timers in the Bristol Bay canning business state that this has been one of the worst years, as regards weather conditions, that they have ever experienced in that district. Gale followed gale, and in the midst of the season there was a continuous storm for 20 days, making it impossible to fish much of the time. Men who have spent many summers at the canneries say they have never before seen Bristol Bay in such a turmoil, large numbers of boats from all the canneries being driven ashore, and a good many destroyed, while many others were driven so high on the beach that they could not be pulled off until the season was over. Several scows and tugs were also beached, and some of the tenders were badly damaged.

Right—*Fine sailing and no fish.*
Below—*Bad winds and tides forced this boat up on the beach. Now comes the job of disentangling the net and picking the fish.*

COURTESY OF PACIFIC FISHERMAN

HARRY HOYER COLLECTION, SAN FRANCISCO MARITIME MUSEUM

COURTESY OF HAROLD BRINDLE

Left—*Fish hitting and everyone working. Italian fishermen worked with the mast up; the "squareheads" worked with it down.* Below—*Fleet of gill-netters being towed home at the end of the season with a ship waiting at ship's anchorage to load and transport the season's pack. Fishermen were paid run money to longshore at the cannery and to load the ship.*

COURTESY OF JOHN GILBERT

COURTESY OF NATIONAL FISHERMAN

COURTESY OF NATIONAL FISHERMAN

Clockwise from top—*Picking a set net, 1959. These fishermen are using a plywood skiff and a 10-hp motor. Originally, cannerymen thought the set net was for the aged and indigent, now set-net sites are open to all. Set-netters' summer shacks at Naknek. This set net ended up on the beach after a blow at Naknek.*

COURTESY OF WILHO ALENIUS

W. H. MEYER COLLECTION, SAN FRANCISCO MARITIME MUSEUM

COURTESY OF NATIONAL FISHERMAN

Left—*Unloading a boat that swamped while loaded in heavy weather.*
Above—*Experienced fishermen mend nets for the coming season.*

JOHN N. COBB COLLECTION, UNIVERSITY OF WASHINGTON LIBRARY

CLYDA SCHOTT GREELY COLLECTION, ALASKA HISTORICAL LIBRARY

Above—*Tender* Warrior *of Bellingham caught in Bering Sea ice.* Left—*The beach seining season at Karluk was usually highlighted with gang fights over locations because seven canneries were located just above the high-water mark. Scrap metal was dumped in front of a competitor's site to tangle the nets, and flashing knives were not uncommon. Photo was probably taken before 1910.*

Opposite and inset—*Beach seining at Karluk River. The Karluk was the greatest producer of red salmon for its size.*

PHOTOS FROM FRED HEICK COLLECTION, SAN FRANCISCO MARITIME MUSEUM

ANTONE LARSEN COLLECTION, SAN FRANCISCO MARITIME MUSEUM

Right—*Steam tug* Togiak *on Nushagak River.*

Left—*No fishing and low tide at Clarks Point, Nushagak River, 1918.*

JOHN N. COBB COLLECTION, UNIVERSITY OF WASHINGTON LIBRARY

Clearing the marine ways in preparation for winter storage. The Polar *------ at Clark's cannery.*

ANTONE LARSEN COLLECTION, SAN FRANCISCO MARITIME MUSEUM

End of the salmon fishing season at Alaska Packers' Association cannery, Naknek, 1906.

JOHN N. COBB COLLECTION,
UNIVERSITY OF WASHINGTON LIBRARY

MILLER HIGH-POWERED REBUILT BOAT ENGINES

FISHERMAN'S FRIEND. 40 Horsepower Engine with Three Speed Forward and Reverse Gear, $175.00.

For Salt Water Service

Equipped with Distillate Carburetors

Force Feed Oiling Systems
Positive Water Pumps and
Backed by an Absolute Guarantee

Don't let old-fashioned notions rob you of your hard-earned cash

PEARNE REVERSE CLUTCH. Especially designed for High-Powered Engines.

Up to 40 Horsepower................$47.50
40 to 60 Horsepower................$65.00

The old style slow speed heavy duty high priced engine for fishing and other heavy duty boats has about seen its day. The more modern, four cycle, higher speed motor is fast taking its place and does the work with less fuel, less noise and vibration and gives better service in every respect. One after another fishermen and work boat operators are learning this and saving themselves lots of money.

WHEN BUCKING A STORM YOU CAN ALWAYS BANK ON A MILLER REBUILT MOTOR

We select for our boat motors engines that are of the highest type of workmanship and design. Our rebuilding process searches out the minutest detail of every working part and makes it as good as new. Our stock is large, our prices low and our engines the best that money can buy.

WRITE FOR THE BOOKLET "BOAT ENGINE FACTS"

GEO. W. MILLER CO., 719 East Pine Street, SEATTLE, WASH.

REPRINTED FROM PACIFIC FISHERMAN. NOVEMBER 1917. P. 40

From: *Pacific Fisherman*, November 1917, p. 40.

THE NEW MODEL N. & S. ENGINES.

The announcement of the new model N. & S. engine has brought a flood of inquiries to the manufacturers of this well-known Pacific Coast engine. The N. & S. Engine Co. are turning out three new models, built in 16-20 h. p., 25-30 h. p. and 35-40 h. p. They have followed the general tendency in Pacific Coast built engines and are making an engine that will no doubt be quickly adopted by Pacific Coast fishermen, workboat and pleasure boat owners. The present design of the N. & S. four-cycle heavy duty motor is massive in construction, equipped with dependable make-and-break ignition, mechanically operated valves, mechanically operated force feed oiling system, large crankshaft with generously proportioned bearings and crank pins, and water-jacketed exhaust manifold. The equipment is complete in every detail.

The company will have its new catalog ready for distribution within a few days.

"UNION" OPEN CROSSHEAD ENGINES

The Only Type Ever Built in Large Powers

80-600 H. P. OPEN CROSSHEAD TYPE

Direct and visible method of lubricating piston pin.

Easy and quick adjustment of piston pin brasses without removing cylinder head and piston.

Piston pin is removed from heat of piston head and cool air is constantly drawn through sides of cylinder and piston, preventing hot piston pin brasses.

Hand-holes and plates are provided so sediment and mud in water jackets can be removed.

Separate heads, cylinders and valve covers. Not necessary to disturb ignition or valve gear to remove cylinder head.

SOME USERS OF UNION OPEN CROSSHEAD ENGINES, From 80 to 600 H.P.

FERRIES, WATER BARGES, ETC.

Oakland, Antioch & Eastern Railway — U. S. Navy Department

OIL BARGES

Standard Oil Co. — Associated Oil Co. — General Petroleum Co.

KELP BARGES

Hercules Powder Co. — Swift & Co. — Simmons Hardware Co.

DREDGING AND CONSTRUCTION

Yuba Construction — North American Dredging Co.

TRANSPORTATION

Meteor Boat Co. — U. S. Navy Department — Merchants Navigation Co. — Vail & Vickers

TRADING SCHOONERS

Atkins, Kroll & Co. — Williams, Dimond & Co. — Papua Government
Central Pacific Cocoanut Plantations Co. — Petersen & Brown — Burns, Philip & Co.

TOW BOATS

Henry Peterson Launch & Tow-Boat Co. — Star & Crescent Boat Co. — Crowley Launch & Tow Boat Co. — W. Q. Wright
Antioch Tow Boat & Barge Co. — Capt. A. T. Johnson — Wilmington Transportation Co. — W. H. Gregg

FISHERIES

A. Paladina — Union Fish Co. — Red Salmon Canning Co.
New England Fish Co. — Pillar Bay Packing Co. — Pacific American Fisheries

The Panama-Pacific International Exposition gave the Union the highest award and pronounced it "The Best Type of Heavy Duty Marine Engine."

UNION GAS ENGINE COMPANY, Oakland, California

REPRINTED FROM PACIFIC FISHERMAN, SEPTEMBER 1917, P. 55

THESE BUSY ONES USE
ZEROLENE

The Standard Oil For Motor Boats

The owners of these boats have found that their engines give better service when lubricated with Zerolene. Being correctly refined from asphalt-base crude, Zerolene has greater endurance under cylinder heat—keeps its lubricating value longer—maintains a better piston seal, and deposits less carbon.

Zerolene keeps down engine expense and makes your fuel go further.

Four grades—light, medium, heavy and heavy-duty. Zerolene and all Standard products can be obtained at our many convenient water front service stations.

STANDARD OIL COMPANY
(CALIFORNIA)

Liberty, of Seattle, a halibut boat, owned by J. Engdol, equipped with a Standard-Corliss engine, 110 horse-power; lubricated with Zerolene Heavy-duty.

Sitka, of Seattle, a halibut boat, owned by J. Johansen, equipped with a Standard-Corliss 110 horsepower; lubricated with Zerolene Heavy-duty.

President, of Anacortes, cannery tender, owned by the Coast Fish Company, equipped with a Standard-Corliss engine, 110 horse-power; lubricated with Zerolene Heavy-duty.

Chickagoff, on the Juneau-Chickagoff run, owned by the Chickagoff Mining Company, equipped with an Atlas engine, 125 horse-power; lubricated with Zerolene Heavy-duty.

Maryland, of Anacortes, tow-boat and cannery tender, owned by F. L. McGill, equipped with Frisco Standard engine, 40 horse-power; lubricated with Zerolene Heavy-duty.

Scandia, of Seattle, a halibut boat, owned by the National Independent Fisheries Company, equipped with a Standard-Corliss engine, 140 horse-power; lubricated with Zerolene Heavy-duty.

Alma, a ferry-boat, owned by Juneau Ferry and Navigation Company, operating between Juneau, Douglas, Treadwell and Thane, equipped with Speedwell engine, 250 horse-power; lubricated with Zerolene Heavy-duty.

Pilot, owned by Townsend Launch and Ferry Company, equipped with Frisco Standard engine, 16 h. p.; lubricated with Zerolene Heavy-duty.

Rothschild, ship-tender, owned by Rothschild & Co., equipped with Frisco Standard engine, 16 h. p.; lubricated with Zerolene Heavy-duty.

Pinto, owned by Townsend Launch and Ferry Co., equipped with Frisco Standard engine, 20 h. p.; lubricated with Zerolene Heavy-duty.

REPRINTED FROM PACIFIC FISHERMAN, OCTOBER 1917, P. 59

REPRINTED FROM PACIFIC FISHERMAN YEAR BOOK, 1915, P. 111

SELDOVIA SALMON CO.'S TENDER GLOVINA.
This Is the Most Recently Constructed of the Cannery Tenders.

PACIFIC AMERICAN FISHERIES TENDER GLENWOOD.

STERN-WHEEL STEAMER W. H. BANCROFT, NORTH-WESTERN FISHERIES CO.

KUIU ISLAND PACKING CO.'S TENDER KUIU QUEEN.

PHOTOS REPRINTED FROM PACIFIC FISHERMAN YEAR BOOK, 1914, P. 22

Trial trip of the tender Leader, *1910. Her bones are now in Leader Creek just east of Naknek village.*

COURTESY OF HAROLD BRINDLE

INSET REPRINTED FROM PACIFIC FISHERMAN, NOVEMBER 1917, P. 52

For genuine hard work in fishing, commercial or pleasure boats, without delay, trouble or complaint, there is no more satisfactory engine made than the

"CLIFT"

5 H. P. to 150 H. P. Open Crosshead and Trunk Piston Types

Let us show you the points of superiority of the "Clift."

This engine is designed and built by practical men after many years of gas engine experience.

CLIFT MOTOR COMPANY

Box 226, Waterway, Bellingham, Wash.

Send to us for literature and information

REPRINTED FROM PACIFIC FISHERMAN, SEPTEMBER 1917, P. 50

Southwark-Harris Valveless Oil Engine

(DIESEL PRINCIPLE)

Photograph of the six-cylinder 225 b. h. p. Southwark-Harris engine now at Seattle to be installed in halibut schooner Comet as soon as alterations to the vessel have been completed

The equipment supplied consists of air starting bottles, injection air bottle, auxiliary engine and compressor (1000 pounds capacity), small service tank, silencer, lubricating oil tank and strainer, wrenches and board, oil cans, interconnecting piping for above, bilge pump, circulating pump and thrust block.

Write Standard Gas Engine Co., San Francisco, Calif., or any of their agents for further particulars. For agents see page 62

SOUTHWARK FOUNDRY & MACHINE COMPANY

430 Washington Ave., Philadelphia

Old Colony Building, Chicago, Ill. Brown-Marx Building, Birmingham, Ala.

All Pacific Coast sales and distribution of our engines is now being handled by THE STANDARD GAS ENGINE CO., of San Francisco, and its well-known sub-agencies and efficient sales organization, assuring our patrons of Real Service in all the principal ports of the Pacific.

REPRINTED FROM PACIFIC FISHERMAN YEAR BOOK, 1916, P. 63

COURTESY OF PACIFIC FISHERMAN

COURTESY OF NATIONAL FISHERMAN

Above—*The Naknek River deepens just below the North Naknek cannery, affording the pickup boats good anchorage although in a 10-knot tide. Extremely long runouts will cause them to leave for deeper water.* Left—*Tenders at anchor in Bristol Bay are waiting to unload the gill-net boats which are fishing.*

COURTESY OF NATIONAL FISHERMAN

COURTESY OF PACIFIC FISHERMAN

Left—*Unloading into the brail with one-tine peughs. Fish are now handled by hand for quality's sake.* Above—*When power was legalized for Bristol Bay, the outboard motor was added to the sailboat without success.*

Right—*Powerboats ready for loading—1954.* Below—*The 1951 vintage powerboats had a wooden windbreak. Later modernization added the glass windshield. Gill-netters found these a far cry from the sailboat.*

COURTESY OF FRANK POMEROY

COURTESY OF NATIONAL FISHERMAN

Left—*The* North Star, *at one time Admiral Byrd's ship, brine-froze salmon in Bristol Bay in 1956-57. She carried 18 gill-net boats. The author managed her in 1957—a very unprofitable year.* Below—*In 1951 the days of the sailboat were numbered, and the powerboat era had begun.*

PHOTOS COURTESY OF NATIONAL FISHERMAN

From: Moser, Jefferson F. "Alaska Salmon Investigations in 1900 and 1901," *The Salmon and Salmon Fisheries of Alaska.* Washington: Government Printing Office, 1902. p. 310.

GENERAL CONDITIONS—THE LAW

In my former report, pages 38-42, the subject of the law was briefly discussed, and, with the experience since obtained, there seems no reason for changing the recommendations then made, but, on the contrary, they should be emphasizd and augmented. My opinion of traps has been given previously and need not be repeated here, except to say that, whatever legislation is effected, there should be no exceptions made.

During the past three years so-called purse seines have grown in favor, particularly in southeastern Alaska, and it is now claimed by those operating them that it is no longer necessary to resort to barricades, as the purse seines take all the fish before they can enter the streams. These seines are simply deep drag seines, square hung, which, in hauling, are pursed by gathering in the foot rope. Around the larger and more prolific streams of southeastern Alaska, such as Hetta, Karta Bay, and Quadra, there were from 25 to 30 of these seines in operation during the season of 1900, one seine following the other in such rapid succession that few fish escaped capture. Purse seines, or seines used as such, should be abolished; in fact, the only fishing apparatus allowed should be gill nets and drag seines; no exceptions should be made. The use of these appliances should be regulated, and competent inspectors appointed to enforce the law and the regulations.

In the small streams examined this year those away from the lines of travel were all found to have evidences of barricading. In some, where the current is not very strong, stakes, planted on the banks and in the stream bed to support a net held to the bottom by bowlders, answered the fishermen's purpose, while in other cases the regular log barricade, referred to in my previous report, was strongly in evidence.

Judging from the appearance of the village at Afognak, and comparing it with other native communities in Alaska, there seems to be little cause for complaint. The village has rather a neat and thrifty appearance, the log and frame houses have a substantial air, there are many gardens growing potatoes and other truck, pigs and poultry seem to thrive, and several families have one or two cows. The children seen in the village appeared bright, clean, strong, and always well clad.

The half-breeds and natives of Afognak, according to their own custom, have three recognized districts for hunting and fishing. The half-breeds of the village fish and hunt the northwestern section of the island, drained by the streams named Malinof and Paramanof; the natives of the village have the southern end of the island, which includes the reservation stream and extends to the western shore of Little Afognak Bay; the inhabitants of the latter place claim the island to the eastward of their settlement, and Marmot Island. These districts are again subdivided among the different families for hunting purposes, while the streams are open to all belonging to the district. These limits are accepted by all the inhabitants, and the intrusion of any alien is considered an abuse of their customs. It is natural, therefore, that they resent the fishing of the Afognak streams by the canneries, nor can they understand how these streams, which belong to them by tribal rights, can be closed by the Government. The situation here is the same as it is in southeastern Alaska—the native sees the value of the fish to the canneryman, and he wants the streams, not for his own use entirely, but to sell the fish to the canneries. In other words, he wants to own the streams and bar out the canneries, which is clearly untenable. To the appeal of the natives has been added that of the white men of the district.

Right—*Small seine boats, like the* Skippy, *below, destined for Kodiak and Prince William Sound.*

PHOTOS COURTESY OF NATIONAL FISHERMAN

From: *Pacific Fisherman,* October 1917, p. 15.

HEARING ON KARLUK RIVER CLOSING.

TO WHOM IT MAY CONCERN: Whereas it has been recommended that Secretary of Commerce limit or prohibit all fishing in Karluk River and lagoon, and tributary waters, Alaska, notice is hereby given under the provisions of Section 6, of the Act of Congress approved June 26, 1906, entitled "An Act for the protection and regulation of the fisheries of Alaska," that a hearing to determine the advisability of limiting or prohibiting fishing operations in the above-named waters will be held in Room 328, Customs House, San Francisco, Cal., on November 15, 1917, at 10 o'clock a. m., at which time and place all persons interested will be heard. H. M. SMITH, Commissioner. Approved: WILLIAM C. REDFIELD, Secretary.

Above—*This freighter is loaded with new gill-net powerboats for Bristol Bay.*

From right, counterclockwise—*"Pursing up." The purse line runs through rings on the bottom edge of the net. Bringing in the purse line draws the bottom together with the top edge or "cork line" which floats on the surface. Photo was taken in 1955, before the Power Block and the net had to be hauled by hand. The man at the winch deftly coils his line clockwise, and as the net comes aboard it will be ready for the next set.*

PHOTOS COURTESY OF NATIONAL FISHERMAN

COURTESY OF RAY KRANTZ

The Power Block, being demonstrated here, made seining with heavy nets much easier. Invented by Mario Puretic, the Power Block was first used in 1955.

Top—*The* Island Queen *hauling with a Power Block.* Below—*The* Janice, *a modern seine boat, operating in Southeastern Alaskan waters in 1976.*

COURTESY OF GEORGE HAMILTON

COURTESY OF GEORGE HAMILTON

The bark Star of Finland *leaving San Francisco for Alaska, circa 1910.*

HALVORSON COLLECTION, SAN FRANCISCO MARITIME MUSEUM

Transportation

SAN FRANCISCO MARITIME MUSEUM

Loading in San Francisco for a new season in Alaska, circa 1914.

Alaska Packers' Association fleet in winter layup at Alameda, California, about 1920. Winter overhaul and refitting of the fleet for the next salmon season kept crews busy.

COURTESY OF JOHN GILBERT

1914.—SALMON FLEET—1914.

Vessel	Rig	Net Tonnage	Company	From	To	Sailed	Returned
Abner Coburn	Ship	1878	Libby, McNeill & Libby	Seattle	Kenai	Mch. 18	Sept. 28
A. B. Carpenter	Steamer	82	Northwestern Fisheries Co.	Seattle	Kenai	April 2	Sept. 22
Afognak	Steamer	37	Alaska Packers' Assn	San Francisco	Wrangell		In Alaska
Akutan	Steamer	69	Alaska-Portland Pkrs. Assn.	Portland	Nushagak	April 24	Aug. 30
Albert	Bark	624	Olson Bros.	San Francisco	Bristol Bay	April 22	Sept. 20
Alitak	Steamer	73	Alaska Packers' Assn	San Francisco	Chignik	April 2	Sept. 24
A. J. Fuller	Ship	1673	N. W. Fisheries Co.	Seattle	Uyak	April 2	Oct. 5
Benj. F. Packard	Ship	2013	Alaska Packers' Assn	San Francisco	Chignik	April 3	Sept. 9
Berlin	Ship	1416	Alaska Portland Pkrs.' Assn	Portland	Nushagak	April 18	Sept. 1
Bohemia	Ship	1528	Alaska Packers' Assn	San Francisco	Naknek	Mch. 24	Sept. 15
Bonita	Gas str.	45	N. Alaska Salmon Co.	San Francisco	Kvichak	April —	Sept. 20
C. A. Thayer	Schooner	390	P. M. Nelson	San Francisco	Kvichak	April 23	Sept. 6
Chas. E. Moody	Ship	1734	N. W. Fisheries Co.	Seattle	Kenai	April 1	Sept. 26
Centennial	Barkentine	1138	Alaska Packers' Assn	San Francisco	Koggiung	April 9	Sept. 20
Chilkat	Steamer	98	Alaska Packers' Assn	San Francisco	Wrangell	April 17	Oct. 11
Dashing Wave	Barge	941	Taku C. & C. S. Co.	Seattle	Taku		Oct. 26
Elwell	Barge	1356	George T. Myers & Co.	Seattle	Chatham	April 18	Sept. 14
Equator	Tug	42	N. W. Fisheries Co.	Seattle	Chignik	Laid up	
Expansion	Gas str.	113	N. Alaska Salmon Co.	San Francisco	Bristol Bay	April 22	Sept. 21
Geo. Curtiss	Bark	1680	N. Alaska Salmon Co.	San Francisco	Bristol Bay	April 15	Sept. 23
Gertie Story	Steamer	36	Alaska Packers' Assn	San Francisco	Alitak	April 9	In Alaska
Emily F. Whitney	Ship	2107	Alaska Salmon Co.	San Francisco	Nushagak	April 14	Sept. 4
Guy C. Goss	Bark	1430	N. W. Fisheries Co.	Seattle	Nushagak	April 15	Aug. —
Geo. F. Haller	Steamer	107	N. Alaska Salmon Co.	San Francisco	Kvichak	April 22	Sept. 20
Glory of the Seas	Barge	1939	Glacier Fisheries Co.	Seattle	S. E. Alaska	Laid up	
Harvester	Bark	716	N. W. Fisheries Co.	Seattle	Dundas	Mch. 30	Oct. 3
Henry Wilson	Schooner	430	N. Alaska Salmon Co.	San Francisco	Bristol Bay	April 21	Sept. 9
Indiana	Ship	1413	Alaska Packers' Assn	Seattle	Nushagak	Mch. 18	Sept. 4
J. D. Peters	Barge	1030	N. W. Fisheries Co.	Seattle	Hunters Bay	April 20	Nov. 5
Kadiak	Steamer	58	Alaska Packers' Assn	San Francisco	Bristol Bay	April 11	Sept. 13
Kanak	Steamer	....	Alaska Packers' Assn	San Francisco	Wrangell		Oct. 11
Karluk	Steamer	8	Alaska Packers' Assn	San Francisco	Cook's Inlet	April 1	Sept. 13
Kvichak	Steamer	610	Alaska Packers' Assn	San Francisco	Nushagak	April 30	Sept. 11
Levi G. Burgess	Bark	1475	Alaska Portland Pkrs.' Assn	Portland	Nushagak	April 16	Sept. 3
Lewellyn J. Morse	Ship	1271	Alaska Packers' Assn	San Francisco	Nushagak	April 11	Sept. 13
Lehua	Steamer	129	Red Salmon C. Co.	San Francisco	Ugashik	April 18	Sept. 6
Lillian	Steamer	....	Alaska Packers' Assn	San Francisco	Bristol Bay		Sept. 13
Lizzie Vance	Schooner	383	Alaska Salmon Co.	San Francisco	Nushagak	April 10	Sept. 5
McLaurin	Ship	1312	L. A. Pedersen	San Francisco	Kvichak	April 16	Sept. 20
North King	Steamer	40	Alaska Portland Pkrs.' Assn	Portland	Nushagak		In Alaska
North Star	Steamer	82	Libby, McNeill & Libby	Seattle	Nushagak	April 11	Sept. 1
Nushagak	Steamer	478	Alaska Packers' Assn	San Francisco	Nushagak	April 30	Sept. 15
Oriental	Bark	1550	N. Alaska Salmon Co.	San Francisco	Nushagak		Sept. 12
Olympic	Bark	1353	N. Alaska Salmon Co.	San Francisco	Koggiung	April 19	Oct. 3
Pactolus	Bark	1564	Naknek Packing Co.	San Francisco	Naknek	April 11	Sept. 10
Philip F. Kelley	Steamer	93	Geo. T. Myers & Co.	Seattle	Chatham	April 18	Sept. 14
Premier	Schooner	292	Alaska Packers' Assn	San Francisco	Nushagak	April 17	Sept. 3
Prosper	Schooner	229	Alaska Packers' Assn	San Francisco	Kvichak	April 25	Sept. 5
Paramita	Bark	1444	L. A. Pedersen	San Francisco	Kvichak	April 17	Wrecked
Reuce	Ship	1601	Col. River Packers' Assn.	Astoria	Chignik	April 1	Sept. 5
Roy Somers	Schooner	298	North Alaska Salmon Co.	San Francisco	Bristol Bay	April 22	Sept. 7
Sannak	Steamer	29	Alaska Packers' Assn	San Francisco	Naknek		In Alaska
Santa Clara	Ship	1453	Alaska Packers' Assn	San Francisco	Kvichak	Mch. 28	Sept. 20
Shelikof	Steamer	101	N. W. Fisheries Co.	Seattle	Uyak	Mch. 30	Oct. 3
Sintram	Ship	1495	Naknek Packing Co.	San Francisco	Naknek	April 10	Sept. 8
Standard	Ship	1461	N. Alaska Salmon Co.	San Francisco	Bristol Bay	April 15	Sept. 8
Star of Alaska	Ship	1554	Alaska Packers' Assn	San Francisco	Chignik	Mch. 31	Sept. 11
Star of Chile	Bark	874	Alaska Packers' Assn	San Francisco	Egegak	April 14	Sept. 6
Star of England	Bark	1705	Alaska Packers' Assn	San Francisco	Kvichak	April 14	Sept. 19
Star of Finland	Bark	1430	Alaska Packers' Assn	San Francisco	Alitak	April 7	Sept. 22
Star of France	Ship	1514	Alaska Packers' Assn	San Francisco	Naknek	April 11	Sept. 6
Star of Greenland	Bark	1974	Alaska Packers' Assn	San Francisco	Wrangell	April 17	Oct. 28
Star of Holland	Bark	1987	Alaska Packers' Assn	San Francisco	Karluk	Mch. 18	Oct. 4
Star of Iceland	Bark	1856	Alaska Packers' Assn	San Francisco	Nushagak	April 12	Sept. 6
Star of India	Bark	1247	Alaska Packers' Assn	San Francisco	Ugashik	April 10	Sept. 6
Star of Italy	Ship	1474	Alaska Packers' Assn	San Francisco	Naknek	April 15	Sept. 8
Star of Lapland	Ship	3006	Alaska Packers' Assn	San Francisco	Naknek	Mch. 15	Sept. 8
Star of Peru	Bark	976	Alaska Packers' Assn	San Francisco	Loring	April 18	Sept. 19
Star of Poland	Ship	2987	Alaska Packers' Assn	San Francisco	Kvichak	April 20	Oct. 21
Star of Russia	Ship	1981	Alaska Packers' Assn	San Francisco	Cooks Inlet	Mch. 25	Sept. 27
Star of Scotland	Bark	2146	Alaska Packers' Assn	San Francisco	Karluk	April 9	Oct. 11
Star of Zealand	Ship	2987	Alaska Packers' Assn	San Francisco	Naknek	April 10	Sept. 6
St. Francis	Ship	1757	Libby, McNeill & Libby	Seattle	Nushagak	April 10	Sept. 5
St. Katherine	Bark	1090	Red Salmon C. Co.	San Francisco	Ugashik	April 11	Sept. 10
St. Nicholas	Ship	1687	Col. River Packers Assn.	Astoria	Nushagak	April 18	Sept. 1
St. Paul	Ship	1824	N. W. Fisheries Co.	Seattle	Orca	Mch. 20	Sept. 5
Tacoma	Ship	1671	Alaska Packers' Assn	San Francisco	Nushagak	April 11	Sept. 7
Thistle	Steamer	56	Alaska Salmon Co.	San Francisco	Wood River	April 24	
Unimak	Steamer	158	Alaska Packers' Assn	San Francisco	Karluk	April 14	Oct. 9
W. B. Flint	Bark	746	Libby, McNeill & Libby	Seattle	Koggiung	April 10	Sept. 5
Windber	Steamer	1820	Pac. Am. Fisheries	Bellingham	King Cove	Various dates.	Various dates
Wm. T. Muir	Steamer	43	Alaska Pacific Fisheries	Seattle	Yes Bay	May 4	Oct. 27

Source: *Pacific Fisherman Year Book*, 1914, p. 46.

Right—*The* Abner Coburn *forced ashore by the ice, 1918, at Cape Menshikof, on north coast of Alaska Peninsula, 16 miles west of Ugashik.*

H. D. GILL COLLECTION, SAN FRANCISCO MARITIME MUSEUM

BERTONCCINI COLLECTION, SAN FRANCISCO MARITIME MUSEUM

Above—*On the* Santa Clara *making repairs. Most seamen were competent riggers in the 1920's.* Right—*Crew members, fishermen and passengers from the* Abner Coburn *cooking on the beach at Cape Menshikof.*

SAN FRANCISCO MARITIME MUSEUM

CAPTAIN S. O. JENSEN COLLECTION, SAN FRANCISCO MARITIME MUSEUM

MRS. WILLIAM PETERSON COLLECTION, SAN FRANCISCO MARITIME MUSEUM

FRED HEICK COLLECTION, SAN FRANCISCO MUSEUM

Top left—*The* Tacoma *foundering in ice in the Bering Sea, May 18, 1918.* Bottom—*Bark* St. Katherine *in ice off Ugashik River, May 16, 1918.* Above—*Alaska Packers' Association fleet at mouth of Nushagak River in early spring.*

Top right—Star of India *waiting to load the salmon pack at Nushagak.* Bottom right—*Tug* Equator *and tender* Alitak *towing* Star of Alaska *into Chignik Bay. The* Star of Alaska *is now the* Balcutha *in San Francisco.*

FRED HEICK COLLECTION, SAN FRANCISCO MARITIME MUSEUM

MRS. MARY JOSEPH COLLECTION, SAN FRANCISCO MARITIME MUSEUM

Above—*The captain's responsibilities were eased by the comfort of his cabin on the* Falls of Clyde. *At the back is Helen Joseph, 5 weeks old, daughter of Captain John F. Joseph and his wife Mary.*

SAN FRANCISCO MARITIME MUSEUM

CLYDA SCHOTT GREELY COLLECTION, ALASKA HISTORICAL LIBRARY

FRED HEICK COLLECTION, SAN FRANCISCO MARITIME MUSEUM

Left—Star of Greenland, *Alaska Packers' fleet, taking in sail while waiting for the tug at Wrangell, where she was station ship from 1914 through 1920 and 1922.* Left below—*Time was well spent in net making on the trip North. Here fishermen work on a hatch of the* Star of Alaska.

COURTESY OF HAROLD BRINDLE

Lucile *on spit, 1908; all hands waiting to be rescued.*

Right—Charles E. Moody *burning at Naknek anchorage, June 28, 1920. The ship was a total loss and cause of fire was not determined. What was left of the hull sank the next morning. She was a fine example of an American-made wooden ship and was owned by North Western Fisheries, Seattle. Photo was taken at night from the bark* Star of Holland.

AXEL WIDERSTROM COLLECTION, SAN FRANCISCO MARITIME MUSEUM

SAN FRANCISCO MARITIME MUSEUM

Above—*The* Jabez Howes *made Chignik harbor and then had trouble.* Right—*The* Sintram *blew ashore at Egegik River May 7, 1915, but the cargo was saved. Built in 1877 at Freeport, Maine, her bones are now inland and covered with tundra.*

CAPTAIN PETER LARSEN COLLECTION, SAN FRANCISCO MARITIME MUSEUM

JOHN N. COBB COLLECTION, UNIVERSITY OF WASHINGTON LIBRARY

SAN FRANCISCO MARITIME MUSEUM

Above—*Wreck of bark* Servia *on Karluk beach. The* Servia *sank on November 6, 1907, and the wreckage washed up on the beach. A ship was sent from San Francisco to pick up the crew members and they arrived home on Christmas Eve.* Left—*The* Star of Falkland *also joined the casualty lists.*

From: *Pacific Fisherman*, May 1917, p. 21.

Disasters in Bristol Bay District

Libby, McNeill & Libby's ship St. Francis, bound from San Francisco for the company's cannery at Koggiung with a full cargo of supplies and nearly 250 cannery workers and fishermen, went on the rocks in Unimak pass on the morning of May 14, and is reported a total loss, together with her cargo. Her radio calls were answered promptly by the tug Goliah, the U. S. coast guard cutter Unalga and the Pacific-American Fisheries' steamer Norwood, which took off the employees and landed them at Dutch harbor. The report of the disaster was transmitted by the Government radio station at Dutch harbor.

The St. Francis left Seattle March 2, with a part cargo of lumber, etc., and took on additional supplies and help at San Francisco, starting for the cannery April 20, in command of Capt. J. A. Rosengren.

Right—*Alaska Packers' Association steamer with fishermen and cannery people. The lifeboat was a little inadequate.* Below—*The Italian fishermen were never at a loss for entertainment. This songfest is aboard the* Mt. Baker *in 1924.*

CAPTAIN PETER LARSEN COLLECTION, SAN FRANCISCO MARITIME MUSEUM

COURTESY OF WALTER YONKER

Right—Star of Scotland *at cannery wharf at Loring, on west coast of* Revillagigedo Island.

JOHN N. COBB COLLECTION, UNIVERSITY OF WASHINGTON LIBRARY

Eight salmon packets beating to enter Unimak Pass. A knowledgeable skipper and favorable tides and winds were required. Photo was taken before 1915.

CAPTAIN ECKHART COLLECTION, SAN FRANCISCO MARITIME MUSEUM

CLYDA SCHOTT GREELY COLLECTION, ALASKA HISTORICAL LIBRARY

Right—*A dance aboard the* Star of Greenland *while at Wrangell.*
Below—*Cannery barge taking crews to sailing ship at the end of the season. Everyone went "first class."*

JOHN OMLI COLLECTION, SAN FRANCISCO MARITIME MUSEUM

The Star of Alaska *loading canned salmon at Naknek, Alaska, for the trip south. Wooden boxes were made by hand prior to the season.*

FRED HEICK COLLECTION, SAN FRANCISCO MARITIME MUSEUM

Pacific Coast S. S. Co.'s Steamer Queen.

The Pacific Coast Steamship Co. owns over 20 steamers which are run to all ports of importance in California, Mexico, Oregon, Washington, British Columbia and Alaska. GOODALL, PERKINS & CO., San Francisco, are the general agents.

SAN FRANCISCO MARITIME MUSEUM

From: *Pacific Fisherman*, November 1917, p. 18.

SALMON SHIPS TAKEN BY THE GOVERNMENT.

Notice was received Nov. 5 that seven sailing vessels of the Alaska Packers Association, ranging from 1,470 to 2,487 tons, have been taken over by the U. S. Shipping Board and will be put on the run between San Francisco and Honolulu. The vessels, it is stated, are the Star of Poland, Star of Alaska, Star of Scotland, Star of Italy, Star of Russia, Star of England and Star of Finland.

COURTESY OF WALTER YONKER

CLARA W. JAMES

Left—*Loading the SS* Santa Cruz. *Alaskan salmon was shipped throughout the world.* Above—*Waiting for the pack.*

COURTESY J. T. LIVESAY

ROBERT WEINSTEIN COLLECTION, SAN FRANCISCO MARITIME MUSEUM

SAN FRANCISCO MARITIME MUSEUM

After the sailing ships came the steamers, serving coastal communities, canneries, salteries and mines. Steamers operating off the Canadian coast and in Southeastern Alaska had to contend with narrow channels and rocky shoals and their captains, running with only a compass and a watch, became legendary. Left above—*SS* Queen *in Glacier Bay.* Left—*The* Mariposa, *wrecked October 8, 1915, at Bella Bella.* Above—*The passenger freighter* Jefferson *loading salmon at Lake Bay, 1909.*

CLARA W. JAMES

From: *Pacific Fisherman,* May 1917, p. 21.

WRECK OF THE "THISTLE."

The wreck of the cannery tender Thistle, owned by the Alaska Salmon Co. of San Francisco, off the coast of British Co umbia at Pender Island, will not seriously inconvenience the operations of the company in the north as already a new tender has been secured. The company was able to purchase the power boat Twilight in Seattle.

The Thistle struck a rock at Pender Island during a heavy rain storm, but the persons aboard were able to get ashore before the vessel slid off the rock and sank. The boat was in command of Captain Kahlin and besides a crew of sixteen were C. P. Hale, president of the company, and Mrs. Hale, en route to the cannery at Nushagak. The wrecked party, after waiting on the beach for nine hours, were picked up by the Admiral Farragut and brought back to Seattle. The Thistle had on board a cargo of coal and some few supplies, but the bulk of the cannery equipment was on board the company's barkentine Emily F. Whitney, which sailed north April 8.

Far left—*Loading cases of red salmon at Bristol Bay.*
Below—*The SS* Humboldt.
The steamers were the only "outside" contact for cannery crews, and so important to the way of life that many people were able to identify individual ships by their distinctive whistles.

SAN FRANCISCO MARITIME MUSEUM

Right—*The* Admiral Evans *was one of the mainstays of the Pacific Steamship Company fleet. She is pictured here in Ketchikan in 1916.*

Below—*The* Culburra, *a wooden coal burner, was a World War I freighter and served herring plants and canneries.*

PHOTOS FROM SEATTLE HISTORICAL SOCIETY

PHOTOS FROM SEATTLE HISTORICAL SOCIETY

Above—*The* Spokane, *later renamed the* Admiral Rogers, *was a favorite with cannery crews. Photo was taken at Wrangell in 1915.* Left—*The last iron-hulled Alaska ship was the* Victoria. *She served in the Boer War and was a well-known Nome supply ship.*

PHOTOS FROM SEATTLE HISTORICAL SOCIETY

Above—*The* Windber *was a charter vessel and is pictured here at King Cove cannery in 1913. The SS* Alki, *a regular passenger and freight steamer on the Southeastern Alaska run, was wrecked on Point August on northeast coast of Chichagof Island in 1915.*

PHOTOS FROM SEATTLE HISTORICAL SOCIETY

The Alameda *(above) and the* Cottage City *(left) were frequent callers in Southeastern Alaska in the 1920's.*

Left—*The seaplane came into use to take cannery crews from community airfields to the canneries. Then roads and small cannery airfields were built and small land planes moved passengers.* Below—*Travel via large plane is a far cry from the early day sailing ship. Within a few hours fishermen can travel several thousand miles from home to the job.*

COURTESY OF HAROLD BRINDLE

COURTESY OF NATIONAL FISHERMAN

Below—*First shipment of Alaska canned salmon by barge. Photo was taken at Pelican in 1954.*

LARRY FREEBURN

Right—*The barge is nudged into the cannery dock for loading as soon as the tide allows. Photo was taken at the Red Salmon Canning Company plant, which burned in 1973.*

COURTESY OF CROWLEY MARITIME CORPORATION

PHOTOS COURTESY OF PACIFIC FISHERMAN

Above—*Power scows, such as this, were contrived by the canneries and shipyards. Many were adapted for wartime use then returned to cannery use at the end of World War II.* Right above—*Newly built power scows are shipped by freighter to Bristol Bay to save the long costly run.* Right—*Scows carry the fish on deck and, due to the short period of high water, fish are sluiced off into the cannery elevator for quick unloading, sometimes at the rate of 1,000 fish per minute.*

PHOTOS COURTESY OF CROWLEY MARITIME CORPORATION

Beginning in 1970 salmon shipped out of Bristol Bay was palletized, thus eliminating individual handling of cases. Pallets of canned salmon, automatically "bright stacked" (without cases), are enclosed in fiber sleeves and shrink-wrapped in heat-applied polyethelene, then barged to a waiting ship. In recent years van ships have replaced the steamers. Above—*Loading a barge.* Left—*A barge loading onto a freighter.*

Right—*Plastic containers are loaded with raw salmon for air shipment.* Far right—*Aluminum totes are used for large shipments of salmon. Totes are loaded onto planes with a forklift.* Below—*Palletized salmon for loading onto barges at the plant is an improvement over the old method of case loading salmon onto small scows, for reloading onto ships.*

COURTESY OF BOB SCOTT

COURTESY OF CROWLEY MARITIME CORPORATION

COURTESY OF CROWLEY MARITIME CORPORATION

Left—*Pallets of "bright stacked" shrink-wrapped salmon headed for a waiting freighter. Each cannery uses a distinctive color for the polyethelene shrink wrap to keep its product separate from those of its competitors.*

COURTESY OF CROWLEY MARITIME CORPORATION

COURTESY OF BOB SCOTT

Left—*Canneries adjacent to fishing grounds are still the main means of processing large quantities of fish, but surpluses can be easily moved to other facilities via large planes. The C-130 shown here carries 44,000 pounds.* Below—*A Hercules being loaded.*

COURTESY OF BOB SCOTT

Alaska Canned Salmon Industry 1879-1973

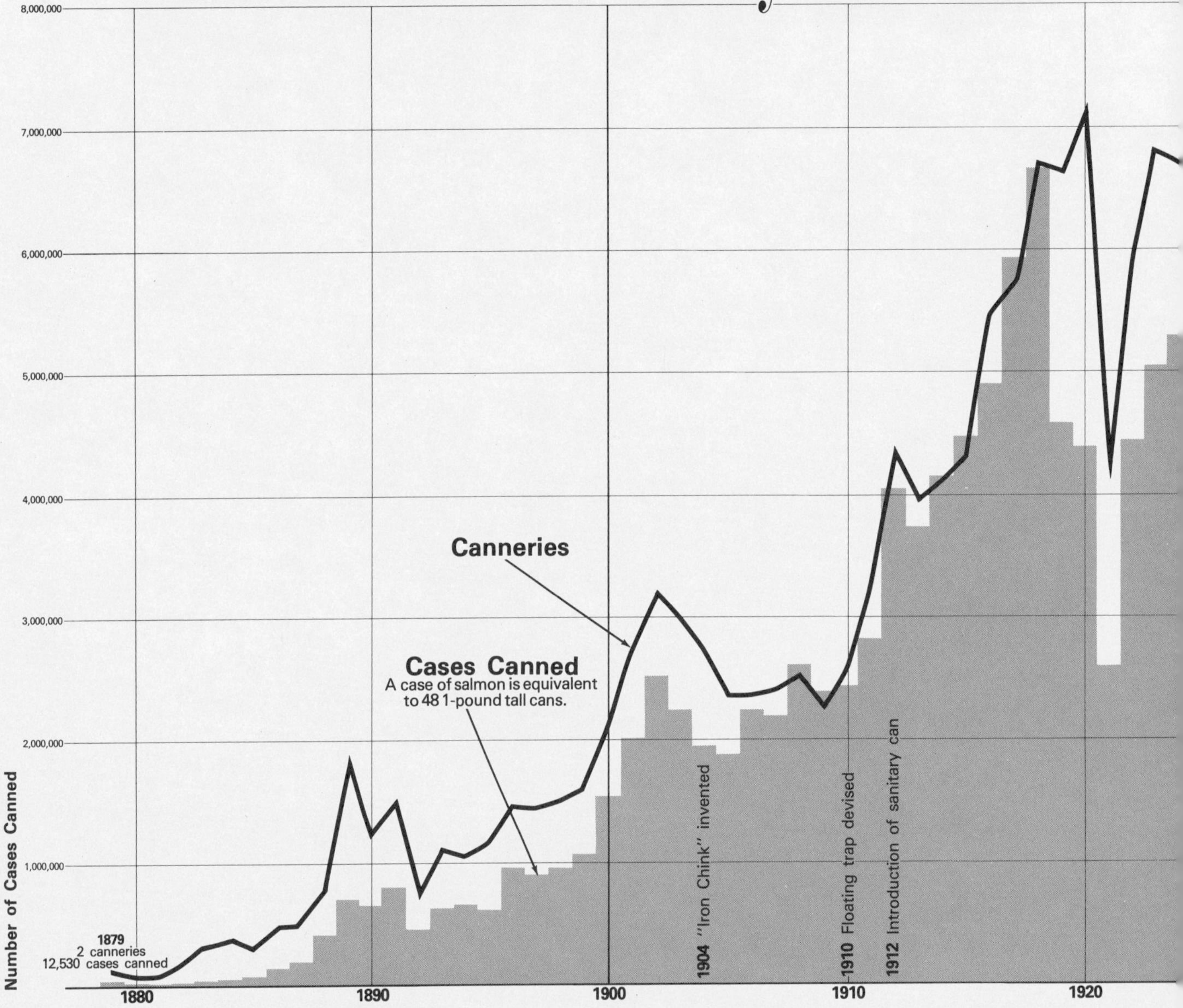

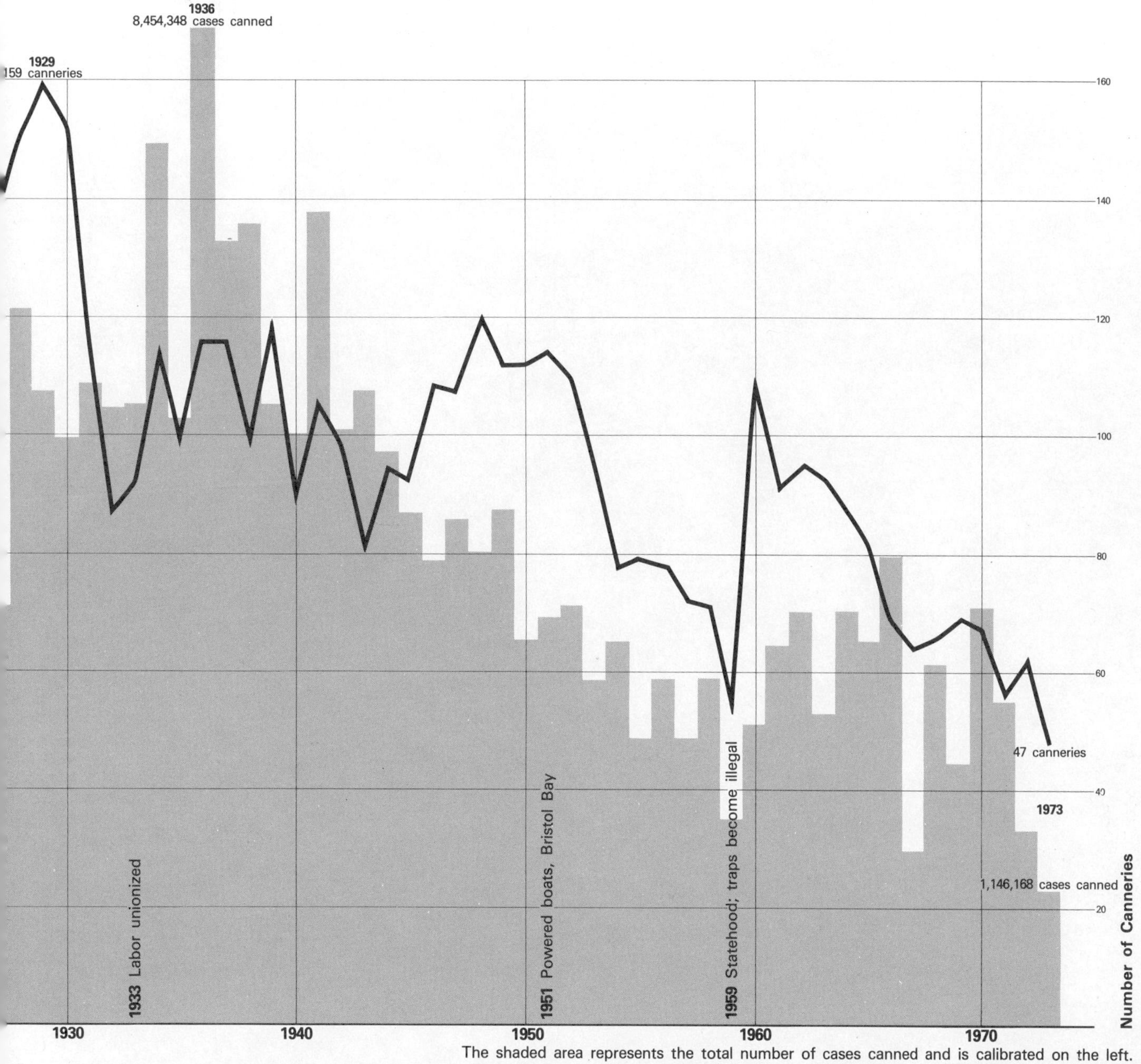

The shaded area represents the total number of cases canned and is calibrated on the left.
The solid line represents the number of canneries operating and is calibrated on the right.

CHART BY JON.HERSH

In the early forties, the owner-operated canneries were on the decline. This is an early day Southeastern Alaska plant, now gone.

COURTESY OF WALTER YONKER

EPILOGUE

CAN THE SILVER YEARS RETURN?

With the decline of the salmon industry since World War II, when consolidation of companies was necessary for the war effort, and with the present decline of salmon runs, one wonders what the future holds for the canning business.

There is an increased demand for fish products throughout the world. To meet this demand, cold storage plants have increased production but such processing is inadequate for large runs of salmon. Modern aircraft and improved methods of transporting fish by air have enabled buyers to process their products at distant points, but again, facilities are inadequate if large salmon runs should develop.

Various state and federal salmon-rearing programs are in the experimental and developmental stages, while the Alaska Department of Fish & Game endeavors to aid Mother Nature to reproduce in her normal bountiful manner despite perhaps-unknown obstacles.

The research efforts by the old federal Bureau of Fisheries and then the state have changed course with the arrival of each new commissioner or director.

Proper action should be taken now, while the industry is hopefully at its lowest economic ebb with only 26 operating canneries, and while the salmon resource is still viable and an invaluable asset to the people of Alaska. All present runs should be properly harvested. At times even now, some few streams are overstocked and not fished.

The 40-year running dogfight between those with industry and fishing experience and those responsible for biological programs should cease. Those charged with managing the resource should be encouraged to seek new approaches and to open communications with the industry and the public to develop programs for salmon runs that could provide future Alaskans with the wealth and employment salmon canning has provided in the past, and the burgeoning appetites of the world with a quality source of protein.

Perhaps this photo history will generate the necessary public concern that there exists a real possibility the fragile salmon runs remaining and the industry they support are in danger of being lost forever.

BIBLIOGRAPHY

Marsh, Millard C. and Cobb, John N. "The Fisheries of Alaska in 1908," Bureau of Fisheries Document No. 645. (Issued April 17, 1909.) *Report of the Commissioner for the Fiscal Year 1908 and Special Papers.* Washington: Government Printing Office, 1910.

Moser, Jefferson F. *The Salmon and Salmon Fisheries of Alaska,* Report of the Operations of the United States Fish Commission Steamer *Albatross* for the Year Ending June 30, 1898. Extracted from U.S. Fish Commission Bulletin for 1898. Pages 1 to 178. Plates 1 to 63. Charts A and B. Washington: Government Printing Office, 1899. (Loaned by Galen Biery.)

Moser, Jefferson F. "Alaska Salmon Investigations in 1900 and 1901." *The Salmon and Salmon Fisheries of Alaska,* Report of the Alaskan Salmon Investigations of the United States Fish Commission Steamer *Albatross* in 1900 and 1901. Extracted from U.S. Fish Commission Bulletin for 1901. Pages 173 to 398 and 399* to 401*. Plates I to XLIV, Plate A, and Charts A, B. Washington: Government Printing Office, 1902.

Pacific Fisherman, A Journal Devoted to the Commercial Fishing Industry. Seattle:
1914, Vol. XII, No. 1. pp. 55, 82.
1917, Vol. XV, Nos. 1-11.
January: pp. 17, 33, 38.
February: pp. 15, 16, 20.
March: pp. 16, 17.
April: pp. 7, 16, 48.
May: p. 21.
July: p. 35.
August: p. 35.
September: pp. 25, 50, 51, 55.
October: pp. 11, 15, 24, 35, 59.
November: pp. 4, 15, 16, 18, 20, 40, 52.

Pacific Fisherman Year Book. Published by Pacific Fisherman, Seattle.
1912: November, Supplement
1914: pp. 22, 35, 38, 46, 51, 57, 62, 63, 75, 78.
1915: pp. 60, 70, 71, 72, 79.
1916: pp. 48, 63, 108.
1917: pp. 50, 52, 53, 67, 81, 94.
1918: p. 88.

ACKNOWLEDGMENTS

We wish to thank the San Francisco Maritime Museum for the use of their photographs including those from the following collections:

Alaska Packers' Association
Martin Astrup
Bertonccini
Bolla
C. W. Brown
Captain Eckhart
H. D. Gill
Halvorson
Fred Heick
Harry Hoyer
Captain S. O. Jensen
Harry Johannesen
J. Johannesen
Carl Johnson
Mrs. Mary Joseph
Antone Larsen
Captain Peter Larsen
Katherine Martin
W. H. Meyer
Mrs. Peter Nelson
S. P. Nielsen
John Omli
William L. Owens
Mrs. William Peterson
Robert Weinstein
Axel Widerstrom

continued

Our appreciation for the use of photographs and other materials from the following institutions, organizations and individuals:

Wilho Alenius
Frank Berry
Galen Biery
Jack Brennan
Harold Brindle
John N. Cobb Collection, University of Washington Library
Crowley Maritime Corporation
Ole Floe
Ted Friele
Jay Gage
John Gilbert
Clyda Schott Greely Collection, Alaska Historical Library
Mrs. Andrew Gronholdt
George Hamilton
Clara W. James
Jon Johnson
Ray Krantz
Bill Lilly
National Fisherman
Jerry Oaksmith
Pacific Fisherman
Frank Pomeroy
Provincial Archives, Victoria, B.C.
Bob Scott
H. C. Scudder
Seattle Historical Society
Vance Sutter
Bob Syre
Bob Thorstenson
Walter Yonker